THE BRITISH ARMY COOK BOOK
1914

THE BRITISH ARMY COOK BOOK 1914

THE WAR OFFICE

AMBERLEY

First published in 1914
This edition published 2014

Amberley Publishing
The Hill, Stroud
Gloucestershire, GL5 4EP

www.amberley-books.com

Copyright © Amberley Publishing, 2014

The right of Amberley Publishing to be identified as the Author
of this work has been asserted in accordance with the
Copyrights, Designs and Patents Act 1988.

All rights reserved. No part of this book may be reprinted
or reproduced or utilised in any form or by any electronic,
mechanical or other means, now known or hereafter invented,
including photocopying and recording, or in any information
storage or retrieval system, without the permission in writing
from the Publishers.

British Library Cataloguing in Publication Data.
A catalogue record for this book is available from the British Library.

ISBN 978 1 4456 4342 7 (print)
ISBN 978 1 4456 4350 2 (ebook)

Typesetting and Origination by Amberley Publishing.
Printed in the UK.

CONTENTS

Introduction	7
General Instructions	16
Apparatus in General Use in the Service	18
Fuel	25
Meat	36
Vegetables and Herbs	41
Beverages	45
Cooking, Various Methods of	47
Recipes	51
Field Instruction	77
Recipes for Field Cooking	130
Appendix	134

INTRODUCTION

The Great War, later to become known as the First World War, was an event that rocked the world. Beginning on 28 July 1914 and ending on 11 November 1918, more than 9 million soldiers and 7 million non-combatants were killed. It was one of the highest casualty rates for any conflict in history, and began a century of social and economic change the world over.

The war involved all of the world's economic powers, which lined up behind the Allies and Central Powers. The Allies initially comprised the United Kingdom, Russia and France, while the core Central Powers were Germany and Austria-Hungary. These alliances changed as other nations were drawn into the war: Japan, Italy and America pitched in on the side of the Allies, and the Ottoman Empire and Bulgaria the Central Powers. By the end of the war, more than 70 million military personnel had taken part in one of the largest wars ever fought.

The war's apparent trigger was the 28 June 1914 assassination of the Archduke Franz Ferdinand of Austria by Yugoslav nationalist Gavrilo Princip in Sarajevo. This set off a diplomatic crisis when the country delivered an ultimatum to Serbia regarding the assassin and alliances formed over the last fifty years were brought to bear. The lines had been drawn, and the fire was soon lit. Within days, the world's nations were at war. The conflict expanded, and it soon spread throughout Europe.

On 28 July, the Austro-Hungarians prepared to invade of Serbia. In response, Russia mobilised, Germany invaded Belgium and Luxembourg and then moved at France, which left Britain no choice but to declare war on Germany. Once the German army's advance on Paris was halted, what later became the Western Front formed, with a trench line that in theory changed very little until over halfway through the war.

On the other side of the continent, on what became

known as the Eastern Front, Russia's forces were successful against the Austro-Hungarians, but were halted in East Prussia by the Germans. In November 1914, the Ottoman Empire also brought troops into the war, forming new fronts in the Caucasus, Mesopotamia and the Sinai. Italy and Bulgaria joined the war in 1915, Romania in 1916, and the US in 1917.

The war moved towards a slow conclusion following the collapse of the Russian government in March 1917, and the revolution in November ended with the Russians coming to terms with the Central Powers. Following this, on 4 November 1918, the

Austro-Hungarian empire signed an armistice. Following a German offensive along the Western Front, later in 1918, the Allies drove the Germans back in a series of successful battles, claiming territory in the trenches. Germany soon agreed to an armistice, on 11 November 1918, finally ending the war.

By the war's conclusion in 1918, four major imperial powers - the German, Russian, Austro-Hungarian and Ottoman empires – had simply ceased to exist, massively changing the balance of economical and political power in the world. Maps of Europe and South-West Asia were completely redrawn, some nations shifting borders and others being disbanded, with several independent nations formed. The League of Nations grew from the ashes of this conflict, given the duty to ensure that nothing of this ilk ever occurred again. As we know, the League failed in this,

and the Second World War would begin barely twenty years after the culmination of the first.

In the words of Napoleon Bonaparte, 'An army moves on its stomach'. The Army School of Cookery was formed in Aldershot in 1885, before the First World War. Previously to this, there had been no official training as such for army cooks. It had simply been the duty of every regiment to provide their own cook, who would ordinarily have come from a cooking background in civilian life. There was no formal training, no central organization, advice or authority on the matter.

It would not be until 1936 that unit cooks were recognized as a speciality, with trade pay introduced and central organization of equipment and stores provided. However, in 1910, the Army School of Cookery, in association with the War Office, printed and published the first 'Manual of Military Cooking'. Rather than being designed for a true cook as we might expect today, this was a guide to army rations, stores, simple recipes and how to cook in bulk for entire units of men, designed to educate the average soldier in how to provide for his fellow Tommies. Unlike other cookery books, there is no assumption of a standard level of skill, and even the most basic of everyday culinary skills – making coffee – is explained in detail. There are also a number of non-standard skills that would never be found in an ordinary cook book, as the army 'cook' in the trenches might be working with less than ideal circumstances or equipment.

As noted in these pages, each battalion had a vat in which food was prepared en masse. Everything from soup to tea would be created within them, which naturally made for some interesting mixtures.

In the first months of the war, soldiers in the British Army were allotted 10oz of meat and 8oz of vegetables per day. This ration was supplemented with

any luxuries from home, including chocolate, cakes etc. Options for meals were limited, but ingenious, as you'll see in the following pages. It should be noted, that although this was the standard ration at the beginning of the war, when this title was published, by the end of the war circumstances were a lot worse. With the German blockade disrupting supply lines and the sheer size of the armies fielded on either sides of every front, rations became scarcer and scarcer. Several years into the war, meat rations were down to 6oz. Flour shortage in 1916 saw dried turnip powder come in as a replacement, and eventually, the standard diet of the British Tommy consisted of pea soup, with a ration of horse meat every nine days.

In terms of ingredients, fresh food would often reach the soldiers on the front too late due to the difficulties and dangers of transport and travel, so a large part of a soldier's diet would consist of dried or packaged ingredients. These would often be mixed up together into a dry and stale mixture. Food was eaten cold in many cases, as the kitchens could not be erected close enough to the trenches for kitchen staff to reach the front quickly enough. There were very few individual camp stoves as the war raged on, as fuel naturally had become scarce along with supplies.

However, despite all these difficulties, propaganda ensured that the British public and their foreign enemies believed the average Tommy to be well fed, healthy and content. That said, this book is a fascinating glimpse, not only into the individual recipes offered to the average soldier in the trenches, such as the much maligned Maconochie stew, pea soup, 'brown' stew and meat pie, but also to the lifestyle and standard practices of the army cook during the First World War.

The British Army Cook Book 1914

food

1 - buy it with thought
2 - cook it with care
3 - use less wheat & meat
4 - buy local foods
5 - serve just enough
6 - use what is left

don't waste it

U.S. FOOD ADMINISTRATION

THE BRITISH ARMY COOK BOOK

GENERAL INSTRUCTIONS.

Kitchens.

Everything connected with the kitchen should be scrupulously clean.

The walls of the kitchen will be swept in the early morning, before they become damp from steam.

The windows will be cleaned at least once a week; during the day they will be kept open at the top, to ventilate the kitchen and to allow the steam to escape.

Duties of Serjeant-Cook and Cooks.

The serjeant-cook will have complete control over the cooks of his regiment or battalion, who should receive their orders from him.

He will detail each cook to the apparatus suitable for preparing the various dishes required for the following day, dividing the work so that each man may know what he has to do, in addition to the cooking for his company.

He will afford every facility for varying the diet of the several messes, so that each company may have a complete change daily throughout the week; and will arrange that the companies using the oven one day, shall have the use of the boilers and steamers the next day, and so on.

He will be personally responsible that no misappropriation of any kind whatever takes place, and should be present when the milk is issued with a list of the quantities ordered, to ensure that each company receives the correct amount.

Groceries should be received by the serjeant-cook, who will weigh each day the quantities of the various articles received for each company, and satisfy himself that they agree with diet sheet and are the correct quantities for the number of men in mess. He will then lock them up in a cupboard, and retain the key. He will issue the various articles to each company-cook, and will see that the full quantity as issued is actually used, and that it is prepared by the cooks according to the instructions given.

When imparting instruction, the serjeant-cook should illustrate his meaning by taking any particular dish and preparing it himself, giving full details during its preparation. When at some future time the same dish is again being prepared, he will see that his previous instructions are carried out, checking errors on the part of the cook. Patience and tact are required, especially with young soldiers, in training them in their duties as cooks.

When assistant cooks are allowed they should be trained under the close supervision of the serjeant-cook with a view to replacing the cooks when required.

The meat when issued to the cook will be at once placed in the dish belonging to the particular mess for which it is intended, care being taken to mark the dish with the number of the mess or room.

When nets are used for vegetables, &c., a tablet or piece of wood, with the number of each mess plainly marked thereon, should be attached to each.

Cooks should not be allowed to have their meals in the cookhouse.

Smoking is not permitted in the kitchens.

CLEANING UTENSILS.

New utensils will be cleaned before they are used.

A new iron pot should first have a handful of sweet hay or grass boiled in it, then be scrubbed with sand and soap; afterwards clean water should be boiled in it for about half an hour. A new tin should be filled with boiling water in which a spoonful of soda has been dissolved, and placed over the fire to simmer; afterwards it should be scoured with soap and rinsed with hot water, the soda renders soluble the resin used in soldering.

Tins can be kept clean by rubbing them gently with sifted-wood ashes. A copper stewpan or vessel can be cleaned with fine sand and salt, in the proportion of half salt to that of sand, then rubbed thoroughly with the hand or a brush. If there be any stains a lemon (or vinegar) may be used to remove them.

Colanders should be well rinsed with boiling water, dried, and the frame cleaned and polished with whiting, care being taken that no particle of dust remains on it before hanging up for future use.

Steamers, dishes and other tin ware should be well washed in soap and soda water and polished with whiting.

Previous to use, all utensils should be thoroughly clean, and, when possible, exposed to the sun daily. The practice of keeping them in cupboards until required for use should be discouraged.

All utensils, after being used, should at once be filled with hot water and placed over the fire to scald thoroughly, then cleaned and well dried.

Grease remaining in a vessel will make it rancid, and moisture will rust it.

In washing any greasy utensil it is better to use the hand instead of flannel, as the latter retains the grease.

Knives and forks (unless plated) should be cleaned with brickdust and flannel, and, if rusty, rubbed with a fresh-cut potato dipped in ashes

Plate or plated articles can usually be kept clean and bright by washing them with soap and boiling water, rubbing them dry whilst hot with soft cloths.

Utensils with bone, ivory or wooden handles should never be placed in hot water.

Large knives, flesh forks, choppers, ladles, bowls, &c., should be well washed with hot water and soda, and afterwards polished with brickdust; they should be at once cleaned after use and put in their proper places in the kitchen.

The meat block and benches should be well scraped, and then scoured with hot water, soap, and soda, and be used for no other purpose whatever, except that for which they are intended.

APPARATUS IN GENERAL USE IN THE SERVICE.

MANAGEMENT OF WARREN'S IMPROVED APPARATUS.

After use the fire should be drawn, and the apparatus allowed to cool down, close the furnace and ashpit doors, then remove the soot cap at the bottom of the stove pipe, insert the flue brush, clean the flue and the top of the oven, sweeping from side to side, close the soot cap and damper in the flue, open the soot caps on top of the oven, sweep the top and sides of it. Close the soot caps, remove the cover under the oven door, clean out the ashes and soot with a rake, sweeping well out with a flue brush, replace the cover, open the ash, pit and furnace doors, clean out the furnace, empty the ashpit, and fill it with water. Lay the fire with ½ a lb. of wood and 7 lbs. of coal. Close the furnace and ashpit doors. Fill the boiler to the gauge tap. In the morning, open the ashpit and furnace door. Light the wood and close the furnace door With a moderate fire the water should boil in 1 hour, the oven be ready for cooking in about 45 minutes, and the hot plate for frying in 30 minutes. When the water boils the coffee should be made, and the boiler refilled and the fire replenished for dinner, closing the damper slightly, until the cooks resume their work after breakfast. Should the cooker supplied for green vegetables or pea soup be required for dinner, it must be placed on immediately after breakfast has been served. When the dinners have been cooked and served, the boilers should be refilled for washing up. After the required quantity of water has been issued for cleaning purposes, the boiler should be refilled again, and the fire banked up for tea.

General Instructions.

1. Avoid the use of the rake; it is only required for cleaning purposes.

2. To economise fuel and ensure the apparatus working satisfactorily it must be cleaned out every day when in use; also keep the space above the bridge clear, and the fire bars free from clinkers.

3. A good fire must be maintained while cooking the dinner; when it requires replenishing, ease the fire with a poker, pushing the live coal to the back, placing the fresh in front, adding not more than 5 lbs. of coal; if more than this quantity is added, it will invariably choke the fire and stop the draught.

4. The cooker in front is intended for soups, rice, vegetables, puddings, hams, porridge, &c., but should never be used for tea or coffee. It may be used for providing an extra supply of hot water.

5. Cinders must not be used till after dinner; then sufficient should have been saved from the day's consumption of coal to prepare the tea.

6. All cookers when not in use should be clean, thoroughly dried, and kept in a dry place ready for use.

7. During the process of cooking stews by steam they should be frequently stirred; the dishes in the oven moved about. When there are no potatoes to be cooked, the cookers can be used for steaming puddings. Care should also be taken that the water from the condensed steam is occasionally drawn off.

8. Should the whistle on the feed pipe indicate that the boiler is empty, it must be refilled at once. If this should happen during the cooking of the dinners, the boiler must be filled with hot water so as not to reduce the pressure of the steam.

Dean's Iron Ovens.

The oven is an iron box, which can be closely shut, with a furnace underneath, surrounded by patent fire lumps, and enclosed in a cast-iron frame.

There are three sizes in use in the service, viz., the 2-dish, rated to cook for 25 men, the 4-dish to cook for 50 men, and the 8-dish to cook for 100 men. After being in use, the oven should be allowed to cool, and then thoroughly cleaned by opening the soot caps and oven dampers, above the folding doors; then insert the flue brush and clean the top of the oven. Close the soot caps and oven dampers, open the doors on top, and remove the back and side soot caps, insert the flue brush and sweep well down the back and sides of the oven, replace the soot caps and close the doors on top; remove the soop caps above the furnace doors, and with a rake clear the soot from the back and sides of the oven, sweeping well out with a flue brush; close the soot caps, open the furnace, sweeping on either side with a cinder brush; empty the ashpit and fill it with water.

If required the following morning, lay the fire with 1 lb. of wood and 10 lbs. of coal; close the furnace and ashpit doors. In the morning open the furnace and ashpit doors, oven dampers and damper in the flue; light the wood and close the furnace door. Each fire should be allowed to burn well down before replenishing, then take the rake and clear the bars, pushing the live coal to the back of the furnace, and adding fresh coal in front. With the 8-dish oven the fire should never be replenished with less than 10 lbs. of coal, the 4-dish 7 lbs. of coal, and the 2-dish 5lbs. of coal. During the time the 8-dish oven is in use, should one side appear to get hotter than the other, the oven damper on the hottest side should be closed until the other side is brought to the same temperature, then opened again.

When the oven is sufficiently heated, which can only be ascertained by experience, it should be damped down, by having a clear fire, free from smoke, closing the ashpit and furnace doors, oven dampers, and damper in the flue, leaving the slide of the ashpit door open.

During the time the oven is in use, the folding doors should be opened as seldom as possible. An iron oven takes one hour to bring to the required heat. Including the fuel necessary to keep it hot while the food is being cooked, it will take 60 lbs. of coal.

Dean's Steel Boilers.

Dean's steel boilers are surrounded by patent fire lumps and fitted in a cast-iron frame. They are rated to cook for 50 men each; one, however, will not cook for 50, though two will cook for 100, as it is impossible to prepare soup and steam potatoes on the same boiler, and reserve a separate one for tea. Each boiler will contain 20 gallons of water; after being in use they should be removed from their bearings and thoroughly cleaned outside and inside, the flue swept as far as can be got at with the flue brush, sweeping round the frame with the cinder brush, cleaning out the furnace, emptying the ashpit, and filling it with water.

If for use the following morning, lay the the fire with 1 lb. of wood and 7 lbs. of coal, replace the boiler in its bearings, place in the required quantity of water for breakfast, put on the lid, and damp the boiler down by closing the ashpit and furnace doors and damper in the flue.

In the morning the fire should be lighted by opening the ashpit and furnace doors, and damper in the flue, then light the wood and close the furnace door. As soon as the water boils the coffee should be made, and the fire drawn and placed under another boiler if required for dinner; if not, the cinders should be placed in the ashpit for use at some future time.

To keep the boiler on the simmer, draw the fire off the fire bars on to the dead lump in front of the furnace door, leaving a few live cinders on the fire bars: close the ashpit door and damper in flue to within 1 inch, leaving the furnace door open. To bring the contents of the boiler to boil again, push the fire back on to the fire bars, close the furnace door, and open the ashpit door and damper in flue.

In steaming over a boiler, it should be three parts full of water and at a sharp boil before the steamer is placed over it.

DEAN'S COMBINED COOKING APPARATUS.

Before lighting the fire, thoroughly rake out any ashes or coal from the flue at the back of the furnace.

The fire damper should be used only for:—

(a) First lighting fire; to be closed after half-an-hour.
(b) Reviving the fire quickly.
(c) Heating water for baths quickly.

Of course the opening of this damper greatly increases the fuel consumption and lessens the heating of the hot plates, ovens and boilers.

The oven damper—right.

Full open for oven; right hand.

Half open for (a) stock pot over oven: (b) hot plate over oven.

The boiler damper.

Controls (a) the boiler on left hand; (b) oven left hand top heat.

The oven damper—left.

Controls the oven left hand bottom heat.

Any of the dampers may be closed, concentrating the whole of the heat from the furnace upon that portion of the apparatus controlled by the open dampers.

The apparatus may be damped down by closing entirely the furnace and ashpit doors and all dampers, leaving one slightly open to allow the smoke to pass away; the fire will then remain in for several hours.

To ensure regularity in the working of the apparatus it must be swept and cleaned out every evening after use. To clean the flues: remove all the steamers and cookers, the left hand boiler and boiler over furnace, the loose hot plates over furnace, six soot doors in the hood, and the soot doors under the oven, open all the dampers.

Commence sweeping from the uppermost soot door above the dampers, down through each flue, passing the dampers to each of four soot doors over the hot plates.

From each soot door over the hot plates sweep well down the backs of ovens, also the boiler flue and back of the centre boiler, into the furnace.

Sweep out all soot from the flues which surround the removed boiler and the tops of the ovens into the furnace, rake out all soot and ashes from the flue under the boiler at the back of the furnace and thoroughly sweep down the right-hand side flue of the large oven.

Thoroughly rake out all the soot and ashes from the furnace, leaving the fire bars perfectly clear, also rake down through the slots in the fire cheeks, forming the sides of furnace to clear the down cast flue.

From the front soot doors under the ovens thoroughly rake out all the soot and ashes each side of the iron flue, breaks and midfeathers, taking care that the flues at the back of the ovens are reached and the under sides of oven bottoms are well scraped.

Sweep, externally, the left-hand boiler and replace, replace the soot doors, clean up the whole of the apparatus, replace the various fittings and the apparatus will be ready for use.

Lay the fire with 1 lb. of wood, 20 lbs. of coal.

It is most essential in any apparatus that the cook should make himself thoroughly conversant with the run of the flues and the action of each damper in the regulation of the heat to each part.

Richmond Cooking Apparatus.

Is composed of two distinct parts. (A) consisting of oven and steam chambers, hot plate and boiler for generating steam, also providing water for tea or coffee. (B) portion consists of soup or vegetable boiler and stock pot. There are two sizes in use. The small apparatus will cook for 50 men. The larger cooker will cook for 150 men.

Management of the (A) *Portion.*

After use this portion of the suite should be cleaned by removing the fire, clearing out the furnace and ash pit, closing the furnace and ash-pit doors, and the damper in the flue. Open the cover over the boiler damper, insert the wire brush and thoroughly clean the tubes of the boiler, sweep the top of the oven with a flue brush, close the boiler damper and cover. Open the right soot cap below the oven doors, insert the flue brush sweeping from right to left, close the cap, open the soot cap on the left, sweeping to the left, close the cap. Open the small soot door at the left hand side of the lower oven cleaning the back of the ovens. Close the door and open the three soot caps at the bottom, and with a rake remove any soot and ashes that may have accumulated at the bottom. Clean out the ovens, replace the gratings, close the doors. Open the doors of the steam chambers, remove the gratings, washing them with a solution of hot water and soda, wiping the sides and bottom of the chambers, replace the gratings, close the doors. Empty the condensing box, thoroughly clean the latter, fill with clean water and replace it. Open the valve on top of the boiler. Fill the boiler with the required quantity of water; this is judged by watching the gauge glass and the indicators on the metal protector. Lay the fire with 1 lb. of wood and 7 lbs. of coal, replace the bullseyes on the hot plate. To light the fire open the furnace and ash pit doors, the boiler damper, and the damper in the flue, light the wood and close the furnace door. When the fire has burned down take the small poker and push the live coals to the back of the furnace, keeping the bridge clear, placing the fresh coals in front, adding not less than 7 lbs. When one or both of the steaming chambers are required for breakfast, immediately the water boils the valve must be closed and the steam forced into the chambers, maintaining a fairly good steam. If, on the other hand, the steamers are not required, the fire should be damped down when the water reaches boiling point by closing the boiler and flue dampers, just allowing sufficient draught to carry off the smoke. As soon as the breakfast has been served, the boiler should be refilled, the fire replenished, and the apparatus damped down. On the resumption of the cooking for dinner, the damper in the flue should be opened and the heat directed round the oven.

When preparing the meat for roasting or baking, the largest joints should be done first and placed in the hottest part of the oven, and as each subsequent joint is ready the first joint must be moved to another part of the oven to make room for the next joint, and so on, in order that each piece of meat may be browned on the outside before being placed finally in the position best suited for it to cook in.

With the larger apparatus it has been found convenient to reserve the left hand steaming chamber for meat and puddings, and the right hand chamber for cooking potatoes and other vegetables.

With the 50 men cooker, the higher portion of the steaming chamber should be reserved for meat and puddings, the lower chamber for vegetables, &c. As soon as the dishes to be cooked by steam are ready, the valve should be closed until the steam gauge registers from 2 to 3 lbs pressure; the steam should then be turned on the chamber in use; maintaining this pressure as far as possible. Vegetables can be placed on the steam according to the time they take to cook. Potatoes should be put in the chamber about 45 minutes before the dinner hour, the whole of the pressure of steam being forced on them by slightly closing the tap of the chamber containing the meat dishes, keeping up a fairly good supply of steam until the potatoes are cooked. During the cooking the overflow box of the condenser must be emptied when necessary.

(B) PORTION.

Management of the Stock Pot and Vegetable Boiler.

This portion of the cooker has been designed for making soups, porridge, boiling rice and green vegetables, also for making stock. The boilers are fitted with wire baskets, the stock pot basket having three compartments as receptacles for the three grades of bones. The two boilers are fixed close together in a cast iron frame and are heated by one fire.

The capacity of the stock pot is 15 gallons, that of the vegetable boiler 25 gallons.

The heat can be directed on either of the boilers, or may be allowed to pass into the flue direct as desired by the cook.

To clean out the flues, &c., the fire must be removed, furnace and ash pits cleaned out. Close the dampers. Open the soot caps in the flue, sweep both sides of the flue well down, close the caps; then open the soot doors on either side of the boiler tops, cleaning the sides of the boiler, raking out the soot with a rake. Lay the fire with 1 lb. of wood and 4 lb. of coal. Close the furnace, open the lids of the boiler, remove the bones from the stock, empty the pot. Wash out with hot water and soda, rinse out with clean warm water, replace the stock, remove the bones from the basket, discard the No. 3 bones, wash the basket, replace the bones. Change the tallies, and put the basket in a cool place. Clean out the vegetable boiler, fill 3 parts with water. In the morning, light the fire, first opening the furnace and ash pit doors and dampers, light the wood and close the furnace doors.

When the fire requires replenishing push the live coals to the back, adding the fresh in front. Close the stock pot damper, directing the heat on the larger boiler. Should the latter not be required for breakfast the dampers may be closed just sufficient to maintain combustion.

BRICK OVEN.

Consists of a brick chamber, generally circular in shape, with a low roof not exceeding 20 inches in height. A furnace, constructed in one side, the flame and heat of which passes through the oven to a flue in the opposite side, causes the chamber to become very hot, the heat being maintained until the food placed in it is cooked.

When required for use, the fire should be laid the previous night with 1 lb. of wood and 15 lbs. to 20 lbs. of coal, and the furnace door closed.

In the morning, open the furnace, ashpit doors, and the oven door, remove the fire block from the mouth of the furnace, close the oven door, open the damper in the flue, light the wood, and close the furnace door.

Each fire should be allowed to burn well down before replenishing, then take a rake and clear the bars, pushing the live coals to the back and adding the fresh in front. The fire should never be replenished with less than 15 lbs. of coal. When the oven is sufficiently heated, which is ascertained by looking through the hole in the oven door, and if the soot is all burnt off the top and sides, and the bricks have a bright red appearance, it is ready. Then open the furnace door and see that the fire is perfectly free from smoke; if it is, close the furnace and ashpit doors, open the oven door, replace the fire block at the mouth of the furnace, and clean out the oven with a damp broom or scuffle, close the oven damper and door for a few minutes to allow the dust to settle.

The hottest part of the oven being near the furnace, the larger and coarser joints should be placed there. After being in 1 hour they should be taken out, turned and replaced in the oven until done.

Too much water should not be placed in the dishes, as the steam tends to lower the heat. After a little practice, the heat can be ascertained by merely taking hold of the handle of the oven door, or by placing the hand in the oven.

The door should be opened as seldom as possible. The time a brick oven takes to heat depends upon its construction and the quality of coal used; as a rule, about 1 hour with 50 lbs. of good Newcastle coal, and 2 hours with 200 lbs. of Scotch coal.

Soyer's Stove.

This consists of a 12 gallon boiler contained in an iron cylinder, at the bottom of which is a small fire-place. It will boil vegetables, puddings, &c., for 50 men; it makes a good stock-pot in the field.

FUEL.

It will be obvious that a careful and economical use of fuel will be necessary, in order to carry out the system of messing as now approved. It has been practically demonstrated that with careful supervision, the regulation allowance is barely sufficient to meet all requirements. The following remarks are issued for information on this subject.

1. The allowance of coal for the cookhouse should be issued daily to the serjeant-cook, and care should be taken that an undue amount of slack is not included.

2. The serjeant-cook will be held responsible for the economical consumption of coal, and it will be his duty to regulate the fires, using no more than are necessary for the cooking required. By consulting the Regimental Diet Return, the serjeant-cook will be able to arrange beforehand how his cooking apparatus can be used to the best advantage.

3. Cinders should be carefully preserved, as in some cases they are as valuable for fuel as coal.

4. The following rules for regulating fires and furnaces should be observed:—

(a) Fires should not be kept burning longer than necessary; for instance, when soup has reached its boiling point, a portion of the fire should be withdrawn, also when the brick ovens are heated to the required pitch, the fire should be at once removed and the food cooked by the stored heat.

(b) After fires have been used, but are required subsequently, they should be banked up by placing damp cinders on them, and the ashpit door and damper closed, leaving only sufficient draught to carry away the smoke, the furnace door being kept open.

(c) In replenishing a fire, the live coal should be pushed to the back of the furnace, the fresh coal being added in front. By so doing, the fresh coal becomes gradually consumed, and the heat of the fire is not reduced.

The fuel usually issued for cooking in the service consists of wood, coal, coke, charcoal, and turf or peat.

Wood in barracks is simply issued for kindling purposes, and the allowance is 1 lb. for each 40 lbs. of coal or coke.

The allowance of coal for Warren's apparatus is 3 lbs. per man per week.

With all other apparatus it is 5 lbs. per man per week.

COAL.

The coal mentioned in the scale is seaborne coal or coal rated as such. 100 lbs. of 2nd quality coal, 80 lbs. of coke, or 1 kish of turf of 20 cubic feet will be considered equivalent to 80 lbs. of 1st quality coal.

COKE.

Coke is coal, the bituminous qualities of which have been extracted by heat in close chambers. A ration of coke is 1 lb., but it is seldom issued for cooking purposes.

CHARCOAL.

Charcoal is wood charred in chambers made as airtight as possible.

TURF OR PEAT.

Turf or peat is a substance of vegetable origin, and, when in a dry state, is issued for kindling purposes—$\frac{1}{100}$ kish being equivalent to 1 lb. of kindling wood. It should not be disturbed while burning.

FIRES.

The fire should be prepared as follows:—Cut the wood into small strips, care being taken that it is quite dry, then place small pieces of coal on each side of the furnace, place half the wood crossways, the ends resting on the coals, the remainder lengthways which will allow the air to pass through, cover with moderate sized pieces of coal and light it at the bottom.

The amount of wood and coal required to lay a fire for each apparatus will be:—

Description.	Wood.	Coal.
	lb.	lbs.
Warren's		4
Dean's, 2-dish		5
,, 4 ,,	$\frac{1}{2}$	7
,, 8 ,,	1	10
Steel Boilers	1	7
Brick oven	1	15

REFUSE.

The refuse must be collected after meals by the orderly men of each mess, and taken to the tubs provided, and on no account is it to be allowed to accumulate in the barrack-rooms. The orderly corporal or N.C.O. in charge of each room is responsible that this is done.

Conditions of sale vary much in the service, the principal ones are—that the contractor be held responsible that all refuse is taken from the barracks at least once every day. The contractor to provide the necessary tubs, &c., to contain the refuse. All articles of refuse to be considered the property of the contractor.

The contractor to be held responsible for any nuisances arising through his neglect of the contract.

The money received for the refuse is credited to the men's messing.

To preserve their sanitary condition, the tubs should be frequently scalded out with water to prevent any smell, and the outside whitewashed.

STOCK-POT.

A stock-pot will be established to provide good soup and gravies. It consists of a cooking utensil, either a boiler or large boiling pot, into which should be placed all available bones, &c., such for example as are collected when the ration meat is cut up, in preparing boned and rolled meat, meat pies, meat puddings, stews and curries. This boiler should be kept gently simmering for 4 to 5 hours daily immediately before its contents are required for use. If the ration meat is properly boned it will provide soup for the men of a battalion daily at a nominal cost of peas, lentils, vegetables, &c.

In order to ensure a constant change in the stock, and that no bones remain longer than 3 days in the pot, the following system should be adhered to. The bones extracted from the meat rations on the first day should be placed in a net with a tally attached before being boiled, the bones of the second and third days should be similarly treated; after the third day the bones boiled upon the first day should be removed, and similarly the bones of the subsequent days, the stock being continually replenished from day to day. The bones should always be removed from the stock before the vegetables and other ingredients are added. They should be carefully drained, placed in a dish, and kept in a cool dry place until required the following morning. Every effort should be made in a regimental cookhouse to reserve special boilers or boiling-pots for making stock, in order that, if possible, the surplus portion of unused stock should be carried on from day to day. This process adds considerably to the strength of the soup made.

The amount of water to be added to the bones in making stock must depend on the quantity of the bones. It must be understood that when the stock is not required for soups, gravies, &c., it should be used in preparing dishes such as curries, stews, meat and sea pies, meat puddings, bakes.

DRIPPING.

Dripping is the oil extracted from the fat of all kinds of meat during the process of cooking, and forms a valuable aid to military cookery. It not only provides all the fat required for —

1. Preparation of all paste;
2. Puddings;
3. Issues in lieu of butter for breakfast and tea;
4. Frying purposes;

but the surplus can be sold and the money credited to the messing fund. Thus a considerable saving is effected in the messing expenses of a unit. Every precaution should be taken by all concerned that no fat is wasted and that the dripping handed into store for issue is clean, sweet and free from moisture and all disagreeable flavour. It should be firm and vary in colour from white to pale straw. If dirty or dark brown in appearance, it has either been indifferently clarified or burnt. Fat that is the least scorched should not be accepted, as it will taint everything it may be mixed with. Dirty dripping must be returned to the cook for thorough clarifying before it is taken over. The ordinary ration of 12 ozs. should yield not less than $\frac{1}{2}$ an oz. of dripping which is obtained as follows, and is divided into the two classes:—

First Class dripping is made from the surplus suet issued with the ration meat.

Second Class dripping is the liquid fat that accumulates on the surface of stocks, stews, bakes, pies, &c., and that which settles on the bottom of the dishes when roasting. This fat must be carefully removed before the dishes leave the kitchen; not only to provide dripping, but to render the food appetising, palatable and easily digested. To prepare the above and make it fit for use, it must be clarified in the following manner:—The liquid fat, when skimmed off the dishes, is put into a dish to cool and harden into a solid cake. It is then cleaned, broken up into pieces, put into a clean dish with about 1 quart of water, placed in an oven or on a hot plate and allowed to boil rapidly, all scum being removed as it rises to the surface. When the water has evaporated and the fat becomes clear, strain it into a clean dish and allow to cool; when firm, turn it out in a solid block and carefully scrape away any particle of dirt or impurity that may be adhering to the bottom. This class of dripping must be kept distinct from the other and only issued for savoury pastry and frying. First Class dripping is made as follows:—

After the surplus suet has been collected, it should be cut up into small pieces, the smaller the better, or passed though a mincer, then put into a dish or pot, barely covered with water, put into an oven or over a fire and allowed to boil rapidly until the water has evaporated and the pieces of fat become a light brown colour, then allowed to partly cool, strained through a colander into a clean dish, allowed to harden, turned out and scraped clean as directed for Second Class dripping.

Care must be taken when cutting away the suet from the meat that only the surplus is removed; for instance, the outside fat must not be taken off unless it be excessive. Should it be necessary to do this, the meat must be weighed to ascertain the loss in fat; if the fat is out of proportion to the lean the matter should be reported.

An account of all dripping saved, issued or sold, should be kept by the quartermaster on the forms issued for this purpose. Issues to companies, for any purpose, should be supported by an entry in the messing book. The serjeant-cook should not keep the daily dripping return, he is only responsible for the amount saved, and on handing it into store must obtain the signature of either the quartermaster or quartermaster-serjeant in his book, for the amount handed over.

Dripping should be issued, as far as possible, according to the following scale:—

For what purpose.	Amount.	Quality.
Tea or breakfast	1 oz. per man	
Plum puddings	4 ozs. to each lb. of flour	
Currant rolls	,,	
Jam rolls	,,	
Currant pudding	,,	
Raisin ,,	,,	First Class.
Date ,,	,,	
Treacle ,,	,,	
,, tarts	,,	
Jam ,,	,,	
Apple or fruit tarts	,,	
Bread puddings	2 ozs. to each lb. of pulp	
Plain suet puddings	6 ozs. to each lb. of flour	
Meat pies	4 ozs. ,,	
Sea ,,	,,	
Meat puddings	,,	
Dumplings	,,	Second Class.
Frying fish, deep frying	about 4 lbs. per company of 60 men.	
,, ,, dry ,,	,, 2 lbs. ,,	
,, liver, without bacon	,, ½ lb. ,,	
,, eggs ,, ,,	,, 1 lb. ,,	
,, ,, with bacon	,, ½ lb. ,,	

QUARTERMASTER

Monthly

Date.	Average No. in Mess.	Dripping saved during Month.		Issued Free to Companies.	
		lbs.	ozs.	lbs.	ozs.
Carried forward from last month ...					

Date   191

Place

(*Appendix B.*)

...... .REGIMENT.

DRIPPING RETURN.

LANCE SHEET.

Surplus Sold.		Value At ___ per lb.			Remaining on Hand.		Remarks
bs.	ozs.	£	s.	d.	lbs.	ozs.	

Quartermaster.

_____ *Regiment.*

QUARTERMASTER'S DA

Date.	No. of Men in Mess.	Total Dripping used. lbs.	ozs	Signature of Serjt.-Cook.	A Coy. lbs.	ozs.	B lbs.	ozs.	C lbs.	ozs.	D lbs.	oz
1												
2												
3												
4												
5												
6												
7												
8												
9												
10												
11												
12												

(Appendix C.)

_____ REGIMENT.

...LY DRIPPING RETURN.

For the Month ending_____ 191__

Issued.												Total issued.		Signature of Quartermaster-Serjt. as to issue.
E		F		G		H		Band.				
lbs.	ozs	lbs.	ozs.	lbs.	ozs.	lbs.	ozs.	lbs.	ozs.	lbs.	ozs.	

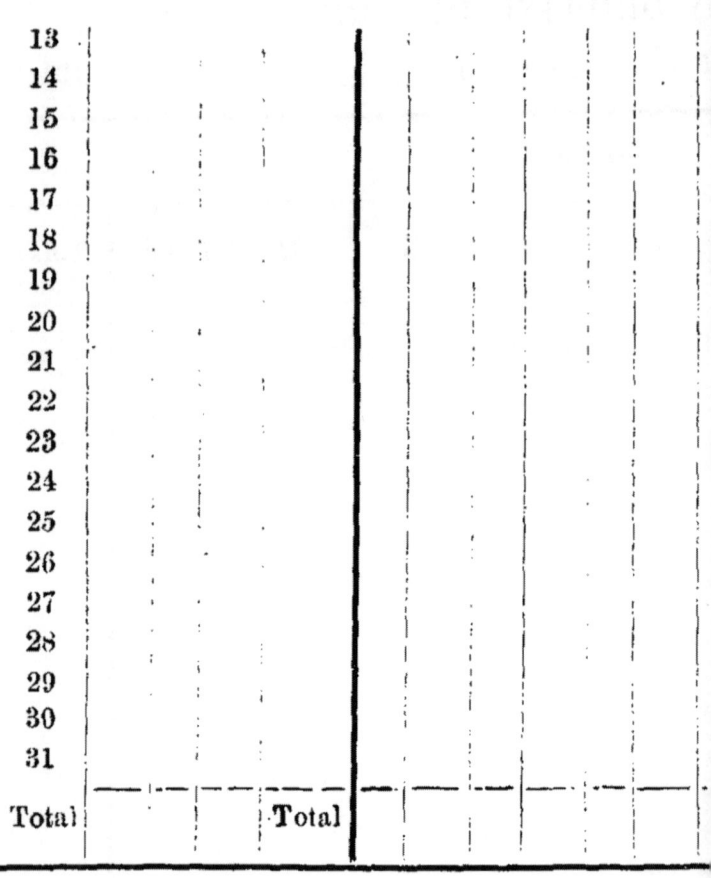

*Place*_____

*Date*_____

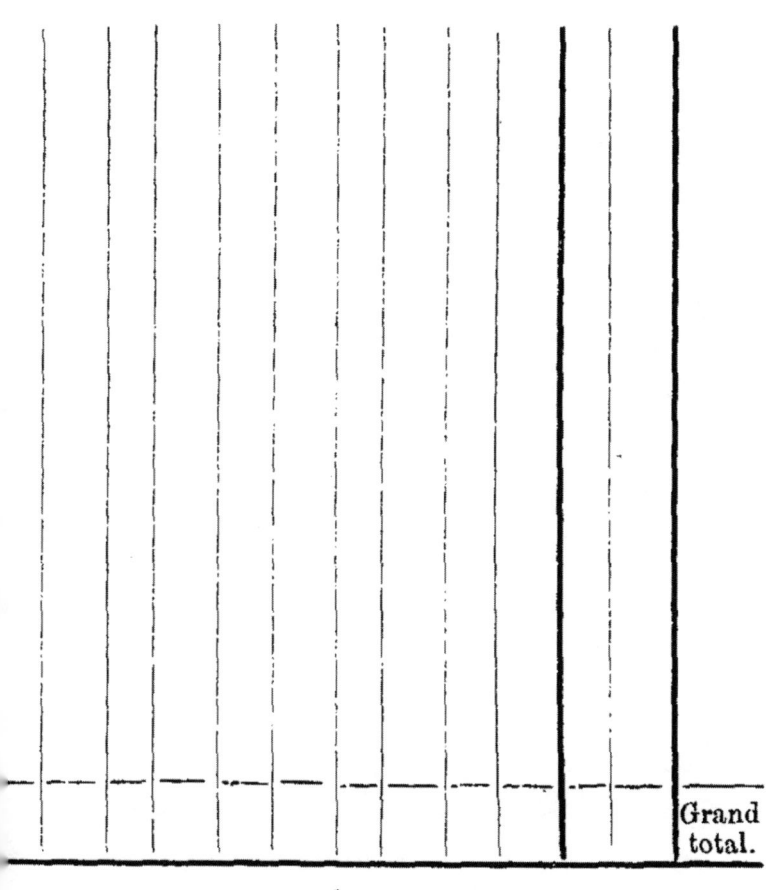

MEAT.

Ration Meat is supplied in the Service under Contract in two ways:—

(i) By the Live Cattle Contract, at stations where Government abattoirs exist, under which the cattle are provided by the Contractor, but slaughtered, and the meat issued by the Army Service Corps. Under this form of contract there is a two-fold inspection, firstly of the live animal, and secondly of the meat after slaughter. Rejections may be made at either inspection.

(ii) By the Meat Contract, at Stations where there is no Government abattoir, under which the meat is supplied ready for issue by the Contractor, and inspected on delivery.

The general conditions of contract governing the class, quality, age, sex, etc., of the meat to be supplied under each system are the same. These are as follows:—

Beef will be supplied to the troops on six days in the week, and mutton for the remaining day.

The beef on four days in each week must be fresh, and frozen beef may be issued on the other two days. The mutton may be frozen.

The right is reserved to issue Preserved Meat from Government stores whenever thought fit. Such issues are limited, as far as possible, to one issue per month, and will be in lieu of fresh Beef.

The meat must be well-fed, good, sound, sweet and wholesome—the beef must be Ox, not under two nor more than five years old, or Heifer and Cow, not under two nor more than four years (48 months) old; the mutton, Wether or Maiden Ewe, not more than four years old. Carcasses of sheep weighing less than 50 lbs. are not to be received.

The term "fresh beef and mutton" means cattle and sheep slaughtered in the United Kingdom, and afterwards neither frozen nor submitted to a longer chilled process than is usual in the trade at Deptford and Birkenhead.

The conditions governing the dressing of fresh meat are as follows:—

Cattle.—Oxen, the root of the pizzle is not to be removed, nor any portion of the cod fat.

Heifers and Cows.—No portion of the udder to be cut away.

Sheep.—Wethers, the pizzle is not to be removed.

Ewes.—No portion of the udder is to be cut away.

Frozen Beef and Mutton is meat slaughtered outside the United Kingdom and imported in a frozen state. The special conditions governing the supply of such meat are as follows:—

Beef.—The quarters are to weigh from 170 to 200 lbs.

Mutton.—Wethers only are to be accepted, carcasses to weigh from 50 to 70 lbs.

Frozen beef and mutton must be thoroughly and carefully thawed prior to issue, and must be in sound condition and free from damage.

General conditions affecting all Meat.—Meat, whether beef or mutton, which is excessively fat, will not be taken unless the contractor consents to remove all surplus fat. In beef, surplus fat is the excessive fat at the kidneys, pelvic cavity, cod fat and udder. In mutton, that on the back and in the region of the kidneys. If the kidneys are removed, the kidney fat must also be taken out.

Delivery.—The ration beef is to be delivered in quarters, the fore and hind quarters alternately, the bone from 4 inches above the knee and upper hock joint to be either excluded, or allowed for in the weight. Mutton is to be delivered in carcasses, excluding heads and shanks from below the knee and hock joints.

Inspection.—In addition to the inspections already referred to, a further inspection may be made by a specially qualified officer, who has power to reject meat already passed. As a result of such an inspection the contractor may be fined within the terms of the contract. In the event of any rejections, the contractor has the right to appeal to the Officer Commanding at the station, and finally to the General Officer Commanding.

Extra ration meat when required, in accordance with the specification, must be supplied by the contractor, to the extent of a quarter of a pound for each soldier included in the ration return, for use in Regimental Recreation Rooms and for other similar purposes, payment for the full amount at contract rates being made monthly by the corps direct to the contractor.

Ration meat when passed, will, if so directed, be cut up, divided and weighed by the contractor, but when cut up by the troops, an additional weight of one per cent. shall be allowed, free of charge by the contractor, to cover losses of cutting up.

The joints in beef and mutton.—Beef :—After slaughter the ox is chopped down, *i e.* divided into two sides. Each side is subsequently divided into two quarters, the divisions taking place between the twelfth and thirteenth ribs.

The usual custom in the service is to "joint" the quarters as follows (*see* plate) :—

In the forequarter :—

1. Clod or sticking piece—five joints of the cervical vertebræ.
2. Chuck rib—three dorsal vertebræ, top ends of three ribs, bottom end of scapula, two cervical vertebræ—should be boned and stuffed, or may be baked and roasted whole. When stuffed, the bones should be made into gravy.
3. Middle rib—four dorsal vertebrae, top ends of four ribs, remainder of scapula. Can be cooked in a similar manner to the "chuck."
4. Fore rib—five dorsal vertebræ, top ends of five ribs. Should be baked or roasted whole.
5. The Plate—lower ends of four ribs. May be boned and stuffed, or stewed, but should not be baked.
6. Brisket—Sternum and lower ends of eight ribs. Is best salted, but may be treated as the "plate."

7. Shoulder, or Leg of Mutton piece. Whole of the humerus, top of radius and ulna.

Should be roasted, baked, or stewed.

8. Shin. Remainder of radius and ulna, less 4 inches, which under terms of contract, must be removed from bottom end. Should be always used for soup or stew.

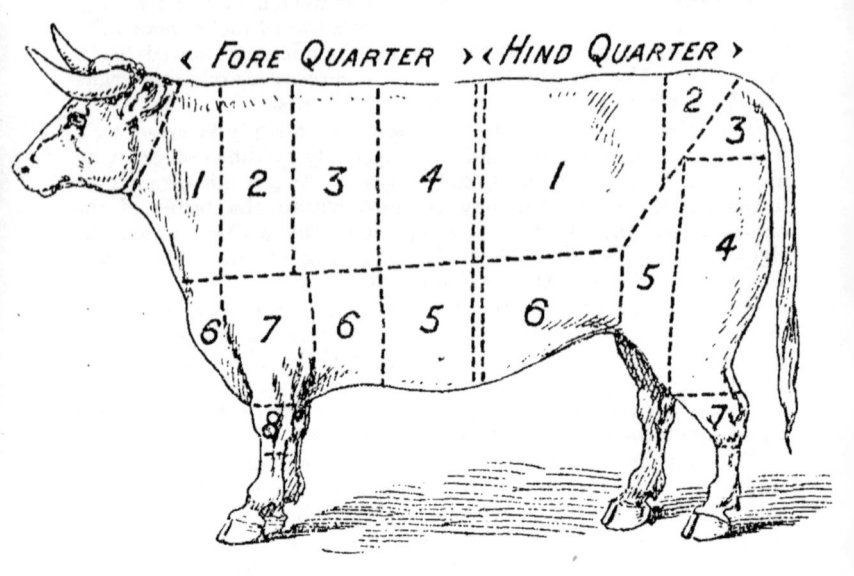

In the hind quarter:—

1. Loin, six lumbar vertebræ, one dorsal vertebræ, top end of one rib and portion of ilium. Should always be roasted or baked.
2. Rump, top part of ilium and sacrum. May be roasted, baked, braized, stewed, or cut into steaks, which may be broiled or fried.
3. Aitch bone, ischium, lower part of ilium, top of femur.
4. Buttock, which is itself divided into two, namely the "top side," inside portion of femur; and the "silver side," outside portion of femur.
5. Thick flank—patella.
6. Thin flank. End of thirteenth rib—should be boned and stewed.
7. Shank, whole of tibia except lower 4 inches, removed under terms of contract. Should always be stewed.

Mutton.—The sheep is not divided into sides or quarters as is the ox.

When cut up for issue, the joints are as follows (*see plate*):—

1. Neck, scrag-end. Should be boiled or stewed.
2. Neck, best end.

3. Shoulder. Should be baked or roasted whole, or the blade bone may be taken out and the meat stuffed and roasted, or it may be cut with the bone in it, and stewed.
4. Breast.
5. Loin. Should be baked or roasted in one piece, or may be cut into chops and stewed.
6. Leg. May be roasted, baked or boiled.

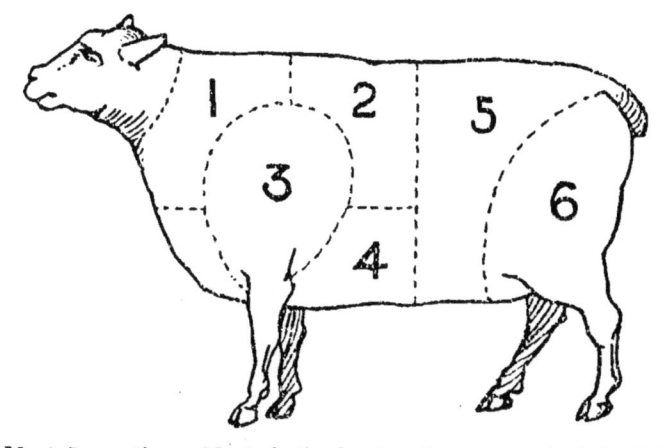

Meat Inspection.—Meat, both beef and mutton, is judged in conjunction with the terms of the current contract, a copy of which should be hung up in every meat store, as regards its age, sex, quality, sweetness and dressing.

Age.—In beef, the carcase externally should have a well-filled and rounded appearance, well covered with bright, clean fat. Internally, the chest and pelvic cavities and kidneys should be well covered with fat; the bones should be ruddy, porous and soft, and plenty of cartilage should be visible on certain of them.

In mutton, the same conditions obtain to a modified degree.

Sex.—In beef the ox, heifer or cow fulfilling the conditions of contract, should be of medium conformation as regards bone and general development, the lean when freshly cut should be bright cherry colour, soft and silky, and well marbled with fat—the fat should be of a biscuit colour, and the crest of medium size.

In the ox, the root of the pizzle should be thin and soft, the erector muscle small, and the cod-fat should be plentiful and lobulated.

In the heifer or young cow, the udder should be a smooth oval pad of solid, or nearly solid fat.

The bull and old cow are not accepted as ration meat. The former—the bull—is distinguished by massive general development of bone and muscle, especially as regards the crest ; absence of fat, dark stringy appearance of lean, with no "marbling," thickness of pizzle and erector muscle, and absence of cod fat. The latter—the old cow—has a generally lean and angular appearance, the bones are white, the flesh is dark and coarse, the pelvic cavity is wide and distended, and the udder is brown, spongy and pendulous.

In mutton, the same characteristics obtain to a rather more modified extent.

Quality—Beef.—A carcase should be healthy and well-fed, and should externally have a well rounded, well-filled appearance. There should be waves of fat on the chest cavity, and plenty of fat on the pelvic cavity and kidneys;—the lean, when freshly cut, should be soft and silky to the touch, full of juice, bright cherry-red in colour and well marbled with fat. The fat itself should be moderately abundant, and usually of a pale straw colour.

The internal organs should be sound and free from disease, and there should be no signs of tuberculous growth or adhesion in the chest and abdominal cavities.

Mutton.—A carcase of mutton should be well-fed and healthy, and should be "mackerel backed"—*i.e.* should have alternate red and white bars over the loins. The fat should be fairly abundant, firm and white.

The flesh should present the same general characteristics as that of beef, except that the "marbling" of fat is seldom present.

Sweetness.—To decide whether meat is sweet or tainted, the senses of taste and smell must be employed. Fresh meat is slightly acid to the taste, while stale meat is distinctly alkaline.

If there is any doubt, the meat should be probed at its thickest portion, with a clean, *wooden* skewer, well thrust in, if possible close to a bone, and the skewer quickly withdrawn and smelt. In beef, the best place to probe a fore and hind quarter is at the chuck rib and pelvic bone respectively, while a carcase of mutton should be cut down between the hind legs, separating the two portions of the pelvic bone.

Dressing.—The method of dressing beef and mutton under the terms of contract has already been alluded to—some of the commoner endeavours to infringe these terms may however be here mentioned.

"Short" forequarters,—*i.e.* those containing only seven ribs, should not be accepted, except in the case of imported meat.

"Stripped" forequarters—*i.e.* those in which any portion of the lining of the chest cavity has been removed, should not be accepted.

If the pizzle, cod fat, erector muscle, udder or crest, has been tampered with, the quarter or carcase concerned should be rejected.

Special characteristics of frozen meat.—The meat is cold to the touch, and particles of ice may be seen on cutting into it with a saw. Its colour is not so bright as that of home-killed meat.

When still frozen, the carcase has, externally, a white appearance, the fat is also white, distinct from the lean, and rather crumbly. There are generally signs of rough handling, and the outside is dirty and untidy.

When thawed, the meat looks sodden, the fat is discoloured, and the exterior of the carcase sweats considerably.

In carcases of frozen mutton, the forelegs are invariably bent towards the body, and as much of the pizzle as can be cut away from the outside, is removed. The conditions of contract as regards dressing do not apply to frozen meat.

VEGETABLES AND HERBS.

The vegetables in common use by the troops are potatoes, carrots, turnips, onions, vegetable marrow, beans, turnip tops, greens and cabbages, and should be prepared as follows.

POTATOES.

Potatoes are best when cooked in their skins, but when it is necessary to peel them, it should be done as thinly as possible, as the best part of the potato is that nearest the skin. After peeling they should be kept in cold water till required for use. If any be spotted in the inside, they should be rejected, as their flavour and the best part of the nutriment has been lost. If for boiling, a little salt should be dissolved in the water before the potatoes are placed in it, but it is better to steam them as their flavour is thereby improved and the waste is less. New potatoes should always be placed in boiling water with a little salt, and not steamed.

Potatoes with rough skins are best for boiling, smooth ones for baking, and as a general rule the smaller the eye of the potato the better is their quality.

CARROTS.

Carrots should always be sent up to table with boiled beef. They vary much in quality, but should be quite firm, and have a crisp appearance when broken. Young carrots should be washed and well scrubbed before cooking; old ones will require scraping and cutting into quarters lengthwise. A little salt should always be boiled with them.

PARSNIPS.

Parsnips, which should be served in a similar manner, are excellent for flavouring, and contain a great amount of nourishment.

TURNIPS.

Turnips are used in all stews, and should be mashed to flavour soups, &c. After boiling, they should be thoroughly drained, a little dripping, pepper, and salt must be added to taste, and they should then be mashed with the ordinary vegetable masher. The turnips should be small, finely grained, juicy, smooth, and sound, and should be peeled, as the part next the skin is fibrous and indigestible.

ONIONS.

The well-known vegetable may be regarded either as a condiment, or as an article of real nourishment. By boiling it is deprived of much of its pungent, volatile oil, and becomes agreeable, mild, and nutritious. As a slight flavouring it is considered an improvement to nearly all made dishes. In stews, pies, &c., it will be found better to first place the onions in a little boiling water with soda, and there allowed to remain for 10 minutes The water, which will then be found quite green, should be thrown away, as it contains the indigestible part of the onion.

Garlic, shalots, chives, and leeks are more pungent than onions, and should be sparingly used

The Spanish onion is larger than the English, and is considered better in flavour.

PEAS AND BEANS.

Peas, beans, and fresh pulse of all kinds should be boiled by placing them in boiling water without salt.

The quicker they are cooked, strained, and served, the more tender they will become.

TURNIP TOPS.

Turnip tops, greens, cabbages, savoys, kale, &c., should be first well picked, washed, and left in salt and water for a short time to drive away any insects that may remain. They should then be placed in plenty of boiling water, with a little salt and soda added, and boiled quickly, leaving the boiler uncovered; this not only helps to preserve their colour, but allows the indigestible part to pass away. As soon as done they will sink to the bottom, and should be taken up at once, strained, and kept warm until served. By so doing they become mellow, and preserve their flavour.

VEGETABLE MARROW.

Vegetable marrows should be peeled, quartered, and the seeds removed (these are very good if added to a soup), they should then be placed in boiling water with a little salt, and boiled until tender. They are also very good mashed, for which they must be boiled, drained thoroughly, and mashed smoothly, adding a little dripping, pepper, and salt to taste.

DRIED PULSE.

Dried pulse, such as the large blue pea, haricots, lentils, &c., should be placed in four times their quantity of water, without salt, and boiled from 2 to 3 hours, until tender, when they will have absorbed all the water, and then seasoned to taste They are a good substitute for potatoes 1 lb. of good dried haricots makes 4 lbs. when cooked. Peas and haricots are a favourite dish when baked or steamed with the meat, for which they must be previously soaked in cold water.

MUSHROOMS.

Mushrooms and marigold flowers are often found growing wild, and give an excellent flavour to a stew or soup.

NETTLES AND SWEET DOCKS.

Nettles and sweet docks are excellent vegetables in the spring, two-thirds of the former being mixed with one-third of the latter. They should be boiled in plenty of boiling water with a little soda. When cooked, drain well, and chop them up as you would spinach, then place them in the dry boiler with some gravy or dripping, salt, and pepper. Stew for about 5 minutes and serve. There are various ways of cooking them, and they are a good substitute for other vegetables in soup.

The young leaf of the mangold wurzel is also excellent when cooked as above. Both should be served with roast meat. Wild sorrel added to pea soup in the spring makes a pleasant change.

SEASONING HERBS, &c.

A faggot of herbs usually consists of 2 sprigs of parsley, 4 of savory, 6 of thyme, and 2 small bay leaves tied together; marjoram may be added. In making soup, the herbs should be sunk by means of a small flint stone. Many of these herbs are found growing wild in this country and in Gibraltar, but it will generally be found advisable to use instead a packet of mixed herbs, at a cost of one penny. A cook should be very careful in detecting the commonest of all herbs, parsley, it being often mistaken for fools parsley, or lesser hemlock, which is of a poisonous nature. This may be detected by bruising the leaves, when they will emit an unpleasant smell, unlike parsley.

CELERY AND MARJORAM.

Celery; the red variety is best for soup, being stronger in flavour.

Marjoram usually grows on chalky soil. Its leaves are small and sharp, and the flower slightly red.

SAVORY AND THYME.

Savory and thyme when not required for use in soups should be dried, powdered and bottled.

MINT.

Mint when dried is used for pea soup. When fresh, it should be chopped up and mixed with vinegar and sugar, which forms a refreshing seasoning for cold lamb.

SAGE.

Sage, a garden plant, should only be used with meats of a strong flavoured, oily nature, such as pork or goose.

Its aromatic qualities promote the digestion of fatty matter.

BEVERAGES.

Tea.

In making tea the vessel must be quite clean, then heated with hot water and rinsed, the dry tea put in, and boiling water poured over it, and the can closely covered for about 8 minutes for the tea to draw.

It should then be strained, and the leaves well rinsed with the additional boiling water required before adding the sugar and milk.

When making large quantities of tea it will be found better to put the dry tea into thin muslin bags, tie loosely so as to allow sufficient space for the leaves to expand and give out their full flavour; put them in the tea vessel, pouring on the boiling water and allow to remain in a warm place closely covered for about 8 minutes; then withdraw the bags, add the milk and sugar, and serve as hot as possible.

Tea should never be made in a vessel that has contained broth or soup.

Coffee.

To prevent adulteration, coffee should be bought in the bean and ground. The beans should be of a bright chestnut brown colour. Care should be taken that only sufficient coffee is ground for the next day's consumption, for when the bean is broken the aroma quickly escapes. Coffee of an inferior quality may be improved by the addition of chicory, but it should not be used in greater quantities than 2 ozs. to 1 lb. of coffee. Beans and chicory are used in adulterating coffee. The presence of the latter may be detected by sprinkling a little of the mixture on some water in a glass. If chicory be present it will at once sink to the bottom, whereas coffee will float for a time. Some should also be shaken up with the water, when the coffee will rise to the surface, and the chicory will sink. In preparing coffee, care should be taken that it is not allowed to boil, as by doing so its aroma is partly dissipated. It should, if possible, be first warmed, which causes each grain of the powder to separate, then the amount of boiling water required should be poured on it. It can easily be prepared in this manner with Warren's apparatus.

With the boilers, the cans should be rinsed with hot water and then the dry coffee placed in them, and the boiling water added gradually, so as to thoroughly extract its strength. It should be made immediately before being required, and served up as hot as possible.

Cocoa.

The concentrated cocoas prepared by well-known firms, such as Fry, Rowntree, Van Houten, form a pleasant change.

Concentrated cocoas should be prepared as follows:—Put the cocoa and sugar into a cup or basin, mix together with a spoon, pour on the boiling water, stirring the whole well, add the milk and serve hot. When preparing it in large quantities, it will be found better to mix the cocoa and sugar into a thin paste with the milk, then add the boiling water, boil the mixture for 3 minutes, keeping it well stirred, then serve. The boiling develops more fully the flavour and aroma of the cocoa.

A teaspoonful of cocoa for each man will be found sufficient; sugar, the same amount as laid down for coffee. Milk as allowed for tea.

CONDIMENTS AND SEASONINGS.

Condiments and spices are salt, pepper, cayenne, mustard, sugar, cloves, allspice, cinnamon, nutmeg, mace, ginger, &c.

SALT.

Salt is almost impossible to adulterate; the finest is known by its whiteness, fine crystallisation character, dryness, complete and clear solution in water.

It is the most important of the condiments, and is used extensively to prevent the decomposition of animal and vegetable substances, and nearly every description of food prepared.

PEPPER.

Black and white pepper is much adulterated with meal, clay, grate rubbish, burnt bread, &c., which tends principally to subdue its strength It should have a pungent aromatic odour, and be hot and acrid to the taste.

CAYENNE PEPPER.

Should be of a bright red colour. It has an acrid, aromatic, and extremely pungent taste, setting the mouth as it were on fire. It is often adulterated with brickdust. The various kinds of pepper are used in soups, stews, &c., to give them a warm biting flavour.

MUSTARD.

Mustard of good quality is known by its sharp acrid taste and smell.

It is adulterated with pea flour, meal, &c., but this is not injurious to health, mustard being too bitter to use by itself. It is used to flavour roast beef, goose, &c., and tends to correct the effect of the strong oily fat in the food, as alluded to when speaking of sage.

SUGAR.

The descriptions usually used in the Service are the white crystal, or granulated—Demerara, and coarse brown sugar. The former is nearly free from adulteration, and is generally used with coffee.

The Demerara is the best description of brown sugar, and should be dry to the touch and not the least sticky or clammy.

Coarse brown sugar frequently contains sugar mites, which may be detected by dissolving a little of the sugar in a glass of tepid water, when they will in a short time float on the surface. It is extensively adulterated with sand, chalk, plaster of Paris, flour, maize, &c.

Sugar is used to preserve fruit, &c., and as a sweetener to many kinds of food and beverages, it is very nutritious.

CLOVES.

Cloves are used to flavour soup, sauces, and puddings.

ALLSPICE.

Allspice for meat, &c.

CINNAMON.

Cinnamon is used in sweet dishes and drinks.

NUTMEGS.

Nutmegs in sweet dishes and various beverages.

MACE AND CURRY POWDER.

Mace and curry powder is used to flavour soups, stews, sauces, &c.

RICE.

Rice varies greatly in quality; Carolina is the best, largest, and most expensive. Patna is almost as good; the grains are small, long, and white; it is used chiefly for curries.

Madras rice is the cheapest and yields plentifully. It forms a most valuable article of farinaceous food, it is light, nourishing, easy of digestion, and cheap, and should be kept closely covered to keep insects from it.

Boiling Rice in small quantities.

(1) Wash the rice in several waters, pick out the discoloured and unhusked grains; and place it on to boil in plenty of cold water. This is the secret of having the rice whole, the water keeping the grains separate; leave it uncovered and bring slowly to the boil;

shake it occasionally to prevent burning, but do not stir it if it can be avoided. When it has simmered gently from 20 to 25 minutes it should be tender. Patna rice will not require quite so long to cook as many of the other varieties. Shake in a little salt, and drain it on a colander, when the grain will separate and be of the finest flavour.

To boil Rice in large quantities.

As it is impossible to cook rice in large quantities in the Deane's and other boilers in use by the above recipe, the following method has been found very good. Prepare the rice for boiling as already directed. Have ready a boiler containing 8 times more water than there is rice, bring to a sharp boil, throw in the rice, draw the fire at once; damp the boiler down, place on the lid, allow to stand for 10 minutes, remove the lid, gently stir the rice, replace the lid, allow to remain for 10 minutes more, see if the rice is cooked, strain off the water and serve. Care should be taken that it is not allowed to remain too long in the boiler; rice should never be overcooked.

COOKING, VARIOUS METHODS OF.

The various methods of cooking in the Service are roasting, baking, boiling, steaming, stewing, frying, and broiling, and may be described as follows:—

ROASTING.

Roasting may be performed in two ways: (1) With an open fireplace; (2) in an oven.

(1) To roast meat is to cook it by exposing it to the direct heat of the fire (360° Fah.). Authorities differ greatly as to the best mode of roasting, but all agree that the fire must be bright and fierce.

To preserve its nutritious matter, the meat should at first be placed close to the fire, basted and slightly dredged, by doing so the exterior of the meat becomes hard, which prevents the escape of the juices; then withdraw it to a greater distance from the fire and baste until done.

The time required is 15 to 18 minutes per 1 lb. weight, but that must be judged by the thickness of the joint and the quality of the meat. Young and fat meat requires longer than old and lean.

If on pressing the lean part with the flat portion of a knife, the meat yields easily, or if the steam from the meat draws towards the fire, it is done.

The loss in roasting is from 30 to 35 per cent., being the greatest average loss in cooking.

Basting is pouring fat or dripping over the meat from a pan placed underneath.

Dredging is shaking over the meat dry flour or bread crumbs and salt mixed.

(2) Joints can only be indifferently roasted however in the Service, and the distinction between what is commonly called a roast and a bake is small. With the former the meat is resting on a trivet (as with a Yorkshire pudding), and to a certain extent the whole surface of the meat is exposed to the dry heat of the oven; with a bake, a certain portion of the meat is resting on the potatoes and becomes partly cooked before being turned. The oven must be thoroughly hot before the dishes are placed in it, and the joint frequently turned, and the dishes moved in the oven so that each portion of the meat may be equally cooked. Baste and dredge frequently. The meat may be either larded or barded previous to cooking; the former is the insertion of pieces of fat in the flesh or on the surface of the meat, the latter is covering lean meat with fat, to impart a richness it would not otherwise possess, both processes are similar to basting.

BAKING.

Baking is a most convenient, economical, and satisfactory mode of cooking certain dishes, such as pastry, meat pies, pork, shoulder of mutton, &c., and it is popular with the troops.

The best oven for baking is one sufficiently ventilated to allow the steam rising from the meat to escape.

In baking pies, &c., they should, at first, be placed in the hottest part of the oven. When the paste is cooked it can be left on the bottom to simmer until done. Pork or other meats of a similar nature, should be covered with a greased paper, to prevent the fat melting too quickly.

The average loss in baking is 25 per cent.

The heat required to bake meat is at least 345° Fahr., but more is necessary should there be much stock or any potatoes in the dish.

BOILING.

The boiling point of water is 212° Fahr., but if salt be added 220° Fahr. will be obtained.

To boil a joint of meat for table, it should be placed in boiling water, and allowed to boil quickly for 10 minutes, then bring it to a simmer and allow it to remain at a temperature varying from 180° to 190° Fahr., it should be surrounded with plenty of water and the lid of the vessel kept on, care being taken that the scum rising to the surface of the water is frequently skimmed off, or the appearance of the meat will be spoilt.

By placing the meat in boiling water the albumen becomes solid and prevents the escape of the juices.

If soup only be required, the meat should be cut into small pieces and placed in cold water and allowed to simmer slowly in order to extract its juices as in preparing beef-tea.

Salt beef or pork should previously be soaked and washed in cold water, then placed in a vessel containing cold water and allowed to boil slowly for one hour, the water in which it has been boiled should then be thrown away. Refil the boiler with fresh cold water, and simmer gently until done. Salt meat requires a little longer time for cooking than fresh. It is placed in cold water in order to extract the salt or brine; if put into boiling water the meat would become hard and indigestible.

The average loss in boiling meat is 15 or 20 per cent., and the time required to cook depends upon its weight and compactness, as a general rule 15 or 20 minutes per 1 lb. weight. If a piece of the flank weighing about 15 lbs. were boiled as issued, it would require about one hour or so, but if boned and rolled it would require from 3 to 4 hours, and would then be sent up to the table as a very substantial joint.

VEGETABLES.

In boiling vegetables to be served separately, (they should be placed in boiling water, with a little salt, and boiled quickly until tender, which will cause them to retain their saccharine juices; but if for soup, they should be placed in cold water and boiled slowly in order to extract the juices.

Cabbages, greens, &c., should be thoroughly cleaned, and allowed to remain for a short time in salt and water, this will destroy any small insects that may remain. They should then be placed in boiling water with a little salt and soda added, and boiled quickly until tender, then taken up and strained and served very hot, the boiler should be kept uncovered, as it not only helps to preserve their colour, but allows the indigestible part to pass away.

PUDDINGS.

Puddings should also be placed in boiling water. The water should be kept at a steady boil. Flour when used as a thickening in soup, requires 30 to 40 minutes' boiling, oatmeal from 50 minutes to 1 hour.

STEWING.

Stewing is considered the most profitable mode of cooking. If properly performed, tough meat is rendered tender and wholesome, and more nourishment is obtained than by any other process of cooking. It should be distinctly understood that stewing is not boiling, all that is required is a gradual simmering, that is, from $170°$ to $180°$ Fahr., and by this process the coarsest and roughest parts of the beast will become soft, tender, and easily digestible.

The best method of stewing is by steam, because with the apparatus now used in Service Kitchens it is nearly impossible to bring the contents of the dishes to boiling point.

Meat of a fibrous and coarse nature, such as legs, briskets, buttock, clods or necks of mutton, should be issued for stewing.

BRAIZING.

Braizing is stewing meat by placing it in an air-tight stewpan, and the heat applied above as well as below, or it can be put in a very hot oven, with a small quantity of water in a dish. Average loss in braizing 10 per cent.

STEAMING.

Steaming is usually performed by steam passing from a close boiler to a close chamber, or by placing a steamer over a boiler containing boiling water, or by placing a few bricks or stones at the bottom of the boiler, covering them with water and placing on them the dish containing the articles required to be cooked.

A steamer is a vessel the bottom of which is perforated with a number of small holes, and should never be placed above a boiler until the water is at a sharp boil.

The articles of food usually cooked by steam are meat, potatoes, puddings, &c.

The average loss in steaming is 15 per cent.

About 1½ pints of water will accumulate from the condensed steam during the process of cooking about 18 rations.

FRYING.

Frying is cooking with the aid of fats, such as butter, lard, dripping, suet, or olive or palm oil, &c., which should be sweet, clean, fresh, and free from salt. With care the same fat should last a long time, but when it acquires a dark brown colour, it is no longer fit for use.

It is customary to place in the pan only sufficient fat to cover the bottom of it, but, when possible, it will be found better to have sufficient fat to cover the article intended to be cooked; in other words, "Frying is simply boiling in fat."

The heat required for frying purposes is from 350° to 400° Fahr. A simple way to ascertain when the fat is sufficiently hot, is to place in it a piece of bread for a few seconds, if this becomes a golden brown colour it is ready; should the bread assume a light yellow it is not hot enough, but if the bread be burnt a dark brown colour, the fat is too hot.

Another way of testing is by spilling a few drops of water into the fat, and when the noise resulting from the evaporation of the water ceases the fat is ready.

The loss in frying depends on the quality of the meat and mode of cooking. If properly performed there should be none.

BROILING.

Broiling is cooking over or in front of a fire, a gridiron being generally used, care should be taken that it is perfectly clean and free from grease. It should be placed on the fire slantways, the lower part in front, this prevents the fat falling into the fire and causing it to smoke. The fire must be clear, bright, and tolerably strong.

Meat for broiling should be cut thin and of an even thickness. The meat then will be equally cooked throughout.

Previous to cooking, the meat should be sprinkled with pepper, but not with salt. It should be frequently turned, and when firm to the touch on being pressed with the flat part of the knife the meat is done.

The average loss in broiling is 8 per cent. With a clean gridiron, a clear fire, close supervision, and the exercise of a little judgment as to when it should be required, small dainty pieces of meat and fish may be cooked by broiling in a manner superior to that obtained by any other process of cooking.

RECIPES.

The advantages of sending everything up to the rooms perfectly hot and well prepared should be impressed upon the cook. No dish looks so unpalatable as a meat pie carelessly made and baked. A pie with a light crust tastefully covered, and baked a nice brown even colour, is in itself an incentive to the appetite. These remarks apply equally to every dish usually prepared in a regimental cookhouse, and the serjeant-cook should satisfy himself that all dishes, &c., leaving the cookhouse are properly served.

GRAVY.

Place the required quantity of stock in a stewpan, bring the contents to the boil; make the thickening by mixing flour with cold water or stock, into a smooth batter; add the thickening, keeping it well stirred to prevent it burning; allow it to simmer gently for 30 minutes; add the bay leaves or mixed herbs and seasoning, and allow the leaves to remain for a few minutes; remove them, colour the gravy by adding a small quantity of caramel made as follows:—

Place 4 ozs. of sugar in a small stewpan, place it on the fire and allow it to remain until the sugar is of a very dark colour, when it will be seen to boil or bubble; add about 1½ pints of water, and simmer for a few minutes; allow it to partly cool and place in a bottle ready for use.

PASTE FOR PIES.

In preparing paste the cook should place his hands under a tap for a few minutes, so that they may be quite cold before touching the ingredients.

Ingredients :—Flour, dripping, salt, and cold water.

Method :—Finely shed the dripping, should it be hard; if soft, it must be broken into small pieces about the size of a walnut. Mix the flour and salt well together, add the dripping, which should not be rubbed into the flour, but carefully mixed. Work the whole lightly into a smooth paste, with the required quantity of cold water, turn out on the table, fold and roll about four times, or press lightly with the hands, roll out, and it is fit for use.

BAKED MEAT WITH POTATOES.

Meat, potatoes, onions, pepper, salt.

Peel and wash the potatoes, having them as nearly as possible all one size; peel, clean, and cut up the onions; grease the bottom of a baking dish, or place in sufficient stock to barely cover the bottom of it; put in the potatoes, leaving a slight hollow in the centre, into which shake the onions, and sprinkle with salt and pepper.

Carefully bone the meat, roll and skewer into a nice joint, place it on the potatoes (worst side up), in order that when the meat is cooked it may be sent to the table with as good an appearance as possible. When it has been in the oven for $1\frac{1}{2}$ hours it should be taken out and turned, then replaced until cooked. Before the dish is sent to the table, all liquid fat should be removed and plenty of gravy poured over the meat. Should the joint be very lean it may be covered with very thin slices of fat before putting in the oven.

BAKED MEAT WITH HARICOT BEANS.

Meat, haricot beans, onions, pepper, salt.

Proceed as for baked meat and peas.

In cooking these dishes, should the peas or beans absorb all the stock, a little more must be added (hot).

BAKED MEAT WITH BLUE PEAS.

Meat, blue peas, onions, pepper, salt.

Soak the peas over-night, and place them in the baking-dish with sufficient stock to cover them; add the onions (sliced up), with salt and pepper; prepare the meat, and proceed as for baked meat and potatoes.

ROAST MEAT AND YORKSHIRE PUDDING.

Meat, flour, milk, pepper, salt, egg powder.

Prepare the meat as for baked meat.

Blend the flour, salt, pepper, and egg powders, well together, in a dish, make a well in the centre, add nearly the whole of the milk, mix into a smooth batter to reduce the lumps of flour, then add the remainder of the milk, beating the batter well up, to make it light, grease the inside of the dish with some liquid fat, pour in the batter. Place on the trivet containing the meat, put the whole into a hot oven. During the cooking the dish must be frequently removed, the meat basted and the batter lifted from the sides of the dish, to prevent it from being burnt.

When the joints are large they should be put in the oven about 45 minutes previous to putting in the batter. This prevents the latter from being cooked before the meat. Care should be observed when putting in the batter that the fat used is clean, sweet and not scorched.

ROAST MEAT STUFFED.

Meat, bread, onions, parsley, eggs, pepper, salt, picked suet.

Crumble the bread into fine crumbs. Free the suet from skin and mince finely. Wash dry and chop the parsley Peel clean and cut up the onions as small as possible, boil them for a few minutes or until they get tender, then strain the water off them. Put the bread crumbs, suet, parsley, and onions into a dish, season with pepper and salt. Mix the whole well together. Break the eggs into a small basin or cup, beat them up well, add to the bread crumbs, etc., mixing thoroughly. When this is done, form the mixture into balls about 2 inches in diameter, and stuff the meat as follows:—

If ribs or sirloin, bone them, lay on the stuffing, roll the meat round and skewer it up, and place it in the baking dish, and allow 12 to 15 minutes per lb. to cook.

Shoulder of Mutton.—Take out the blade bone, lay on the stuffing, and roll up the meat.

Leg of Mutton.—Make an incision above the pelvic bone and fill with stuffing.

The flavour of the stuffing may be varied by using sage, thyme, or majoram. Sage should always be used for goose, duck, or pork.

BROWN CURRY STEW AND RICE.

Ingredients as for a plain stew with the addition of 5 ozs. curry powder and 6 lbs. of rice. *See p. 36.*

MEAT PIE.

Meat, flour, dripping, pepper, salt, onions. Make the paste, separate the meat from the bones, and cut into pieces about 1 oz. each; place a little stock in the baking dish, then the meat and sliced onions, season with salt and pepper, and barely cover with stock, level the surface of the pie, line the sides of the dish with the rough portion of the paste, and cover with the remainder; make a hole in the centre, which will allow the unwholesome gas generated by the confined meat cooking to escape; place in the oven; when the crust becomes firm the pie should only be allowed to simmer gently until cooked; cover with a greased paper if necessary, to prevent the crust being scorched.

STEWS.

BROWN STEW.

Ingredients same as for plain stew, and prepared in a similar manner, and cook in the oven. It requires to be frequently stirred, care being taken that the ovens are only moderately hot.

PLAIN STEW.

Meat, mixed vegetables, onions, flour, pepper, salt.

Peel or scrape, clean and cut up the vegetables and onions; separate the meat from the bones, and cut it, against the grain, into pieces of 2 ozs. each; mix the dry flour, salt, and pepper well together; place a little stock in the cooking vessel, rub the pieces of meat in the dry flour and add to the stock, put in the vegetables and onions, barely covering the whole with stock; let it simmer gently for 3 hours, keeping the vessel closely covered until done.

IRISH STEW.

Meat, potatoes, onions, pepper, salt, stock.

Peel, wash, and slice the potatoes; peel, clean, and cut up the onions; separate the meat from the bones, and cut into small pieces; place a little stock in the cooking vessel, and a layer of potatoes at the bottom, then a layer of meat and onions; season with pepper and salt, then another layer of potatoes, and so on alternately, until the vessel is nearly full, potatoes forming the top layer; barely cover the whole with stock, and stew gently for 3 hours, keeping the vessel closely covered, care being taken that it does not burn.

The surplus fat must always be removed previous to cooking, as an Irish stew should not be greasy.

When preparing stews in camp kettles, care should be taken that they are not allowed to boil, or the meat will become hard and indigestible.

CURRY STEW.

Ingredients the same as for stew, with the addition of curry powder.

Mix the curry with the dry flour, and proceed as for plain stew.

STEAMED MEAT.

Meat, mixed vegetables, onions, pepper, salt, stock, as required.

Peel or scrape, clean and cut up the vegetables and onions. Bone, roll, and skewer the meat into small joints.

Put the vegetables and onions at the bottom of the cooking vessel, season with pepper and salt, barely cover with stock. Put in the meat worst side up. Place on the steam, after cooking for 1 hour the meat should be turned, then replaced on the steam till done.

With a moderate steam will take 3 hours to cook.

STEAMED MEAT WITH PEAS.

Meat, onions, blue peas, pepper, salt, stock as required.

Soak the peas overnight, peel and slice the onions. Bone, roll, and skewer the meat into small joints, place the peas at the bottom of the cooking vessel, over which sprinkle the sliced onions, season with pepper and salt, barely cover with stock, put in the meat worst side up and place on the steam. After cooking for 1 hour the meat should be turned, then replaced on the steam till done.

With a moderate steam it will take 3 hours to cook.

STEAMED MEAT WITH HARICOT BEANS.

Meat, onions, haricot beans.

Pepper, salt, stock as required.

Proceed as for "Steamed Meat and Peas," using haricot beans instead of peas. If during the process of cooking the haricot beans absorb all the moisture, a little hot stock should be added.

MEAT PUDDINGS.

Meat, flour, dripping, onions, baking powder, pepper, salt.

Prepare the paste and divide it into portions; separate the meat from the bones and cut it into thin slices; peel, clean, and slice the onions; have some basins ready, and equally divide the meat, onions, pepper, and salt; barely cover the whole with stock, damp the edges of the basins, and cover with the paste; place eight at the bottom of the steamer, then put in two trivets with a strainer, which will hold eight more basins, and so on until the steamer is full.

With a good steam they will require 3 hours to cook. Each basin should contain two men's rations.

SEA PIE.

Meat, potatoes, mixed vegetables, onions, flour, suet or dripping, pepper, salt.

Peel or scrape, clean and cut up the vegetables and onions; peel, wash, and slice the potatoes in halves lengthwise.

Make the paste, separate the meat from the bones, and cut into small pieces; place some stock in the cooking vessel, add the meat with the potatoes, vegetables, onions, &c., season with pepper and salt, barely covering the whole with stock; cover with the paste, making a hole in the centre. With a moderate steam, it requires 3 hours to cook.

TURKISH PILLAU.

Meat, rice, cayenne pepper, onions, salt, sweet herbs, flour, stock as required.

Peel and slice the onions, separate the meat from the bones, cut it into small pieces.

Place the flour, cayenne pepper and salt in a dish, well mix together; rub the pieces of meat in dry flour. Pour a little stock in the bottom of the dish, add the meat, onions, and sweet herbs, barely cover with stock, and stir the whole together.

Place in the oven, care being taken that it is only moderately hot and the stew frequently stirred.

Boil the rice as directed for boiled rice, having it ready about 45 minutes before the dinners are served.

Line the sides of a separate dish with the rice, pour the pillau in the centre, then replace in the oven till time to serve.

Toad-in-the-Hole.

Meat, onions, flour, egg powder, milk, pepper, salt, stock as required.

Separate the meat from the bones, cut it into small square pieces
Peel, clean and cut up the onions.

Make the batter as directed for Yorkshire pudding.

Put ⅛ of the flour into a dish, add the pepper and salt, mix well together. Rub the pieces of meat in the dry flour, barely cover the bottom of the cooking vessel with stock, put in the meat, onions, and sufficient stock to barely cover the meat, etc. Stir the whole well up, and level the surface. Put it in the oven, allow to stew for 1 hour, remove the dish, pour the batter over the meat, so as to cover it, replace it in the hottest part of the oven; when nicely browned it should be placed on the bottom to simmer gently till done.

Note.—In preparing the batter care must be taken that it is not made too thin, or when poured over the meat, it will sink to the bottom, forcing the gravy to the top.

Beef Rissoles.

Mince finely 1 lb. of Australian beef or mutton, add ½ lb. of bread crumbs; mix thoroughly with it ¼ lb. of dripping, a little salt, rather plentiful supply of pepper. Divide into balls or cones, and fry in boiling fat until nicely browned; serve in a dish with some thick gravy poured round

Beef Rissoles (another way).

Mince finely 1 lb. of Australian beef or mutton, add ½ lb. of bread crumbs; mix thoroughly with it ¼ lb. of dripping, a little salt and pepper.

Place a small piece of butter, about the size of a walnut, in a saucepan, and one onion chopped fine, and stew gently till of a nice brown colour, keeping the saucepan closely covered; add the other ingredients, stir the whole well together, divide into balls or cones, and fry a nice brown colour, and serve with some flavoured gravy.

BEEF FRITTERS.

¼ lb. of Australian beef or mutton, ¾ lb. of flour, ½ pint of water, 2 ozs. of butter, the whites of two eggs.

Make a smooth batter with the flour and water; stir in 2 ozs. of butter, which must be melted, but not oiled, and, just before it is to be used, add the whites of two well whisked eggs. Should the latter be too thick more water must be added. Pare down the beef into thin shreds, season with pepper and salt, mix it with the batter. Drop a small quantity at a time into a pan of boiling fat, and fry from 7 to 10 minutes, according to the size. When done on one side, turn and brown them on the other; let them dry for a minute or two and serve.

A small quantity of finely minced onions mixed with the batter is an improvement.

MINCED BEEF

1 lb. of Australian beef or mutton, 1 oz. of butter, 1 onion, 1 oz. of flour, stock, salt and pepper to taste.

Put into a stewpan the butter, with an onion chopped fine, and the flour; stir these ingredients over the fire until the onion is a rich brown, add the required quantity of stock, season with pepper and salt. Cut, but do not chop the meat very fine, add it to the gravy, stir till quite hot and serve.

Make a border of mashed potatoes round the dish, pour the meat in the centre.

TOMATO STEW.

Ingredients.—Meat 45 lbs., onions 2 lbs., tomatoes 6 lbs. (or 3 tins of preserved tomatoes), flour 2 lbs., salt 2 ozs., pepper 1 oz.

Cut up the meat as for stew. Place the flour, salt and pepper in a dish, well mix together.

Peel and cut up the onions.

Put the tomatoes, onions, and a little stock in the bottom of the cooking vessel, rub the meat in the dry flour, place in the dish, add sufficient stock to barely cover the meat. Stir well. Cook either in the oven, on the steam or over a fire for 3 hours.

BEEF SAUSAGES.

Ingredients for 60 *men.*—9 lbs. lean beef (exclusive of bone, fat or sinew), 3 lbs. suet, 3 lbs. bread crumbs, ½ packet of thyme or majoram, 2 oz. salt, ¼ oz. pepper.

Method.—Cut up the meat and suet into small cubes and pass it through the mincing machine, moisten the bread crumbs until they become a stiff pulp, add the salt, pepper, and herbs to the pulp and mix very thoroughly together, then add to the minced meat and again mix thoroughly together until the ingredients are equally distributed. Soak the sausage skins in cold water for some hours previous to using them. Take the cutter out of the mincing machine, fit on the sausage filler, run a skin on to it and fill with the mixture. To make sausages weighing 8 to the pound, they should on no account be larger than the small end of the sausage filler.

BEEF RISSOLES.

Ingredients for 60 *men.*—18 lbs. lean beef (exclusive of bone, fat, or sinews), 5 lbs. suet, 5 lbs. bread crumbs, 2 lbs. flour, 2 lbs. onions, 1 packet mixed herbs, 4 ozs. salt, ½ oz. pepper.

Method.—Cut up the meat and suet into small cubes, place the flour, pepper, salt, and herbs into a mixing bowl and well mix together, add the meat and suet and rub into the flour, &c., then pass the whole through a mincing machine.

Mix the bread crumbs into a stiff pulp with cold water, add to the other ingredients and mix well until they are equally distributed. Flour a table and weigh out portions at ½ lb. per man; mould into shape and place in a greased dish.

The rissoles should be cooked in a moderate oven for about one hour, then the fat poured off and a rich brown gravy added to them in sufficient quantities to nearly cover; replace in the oven and simmer for another ¾ hour.

BRAWN.

Ingredients.—1 bullock's head, 1 set of cow heels, 1 packet of spice, 1 oz. saltpetre, 2 lbs. of salt, 1½ oz. of pepper, 2 gallons of water if boiled, 1 if steamed.

Trim the head, removing the skin round the outside of the mouth, also the nostrils, and well wash in cold water; then saw it in two lengthways and through the lower part of the cheek bone; also through the thick part of the skull (avoid the use of the chopper, as it is nearly impossible afterwards to remove the small splinters from the bones), thus leaving the head in six pieces. Scald and clean the heels, splitting them lengthways, removing the large bone above the fetlock, place the whole in cold water in which 1 lb. of salt and 1 oz. of saltpetre has been dissolved and allow it to remain for 10 or 12 hours. Place 2 gallons of cold water in the boiler, add the head, heels, and bones, and as soon as it boils allow it to simmer gently for 5 or 6 hours, the fat or scum being frequently skimmed off; remove the bones from the liquid, and if necessary chop or mince any large pieces of meat that may remain, add the remaining salt, pepper, and spice, stir the whole well together, seeing that the meat is equally mixed with the liquor. Dish up in basins, baking dishes, &c., and allow it to cool. If the heels be tough or old, they may be simmered for 1 hour before adding the head.

Cost 5s.   Result—38 to 40 lbs. Brawn.

	s.	d.		s.	d.
Cost of above ingredients	4	9	Sale of 38 to 40 lbs. brawn at 1½d.	4	9½
Coal	0	3	Sale of 2 lbs. dripping at 4d.	0	8
Balance credit	0	8½	Sale of bones	0	3
Total	5	8½	Total	5	8½

Faggots Baked.

Mix a mince meat of calf's liver, or, if more convenient, pig's liver and fresh fat pork. Chop very finely 1½ lbs. of fresh fat pork. Season with onion, sage, thyme, salt, and pepper.

Steam over boiling water, and throw off all the fat. When cold add a large cupful of bread crumbs; mix all well together. Thoroughly flavour with nutmeg, and made up into round balls, which may be baked in a buttered dish, with a small quantity of good gravy, or, as is often done, wrapped separately in a piece of pig's caul. In either case they should be of a pale brown, and cooked very slowly. Time to steam mince meat, half-an-hour; to bake in a moderately hot oven, 45 minutes.

Australian Meat Curried.

Curry, flour, onions, dripping, salt, pepper, and rice.

Place the dripping, curry, flour, onions, chopped up fine in a stewpan; stir the whole together in the stewpan till the onions are nicely browned, add the stock and stir gently till it becomes smooth; let the whole simmer gently for 30 minutes, removing the fat.

Cut the meat up into dices, about the size of a walnut, add this to the gravy, season with pepper and salt; set by the side of the fire till thoroughly heated through.

Make a border of boiled rice round the dish, pour the curry in the centre, and serve hot.

Broiled Beef Steak.

As the success of a good broil so much depends on the state of the fire, see that it is bright and clear, and perfectly free from smoke, and do not add any fresh fuel just before the gridiron is to be used. Sprinkle a little salt on the fire, put on the gridiron for a few minutes to get thoroughly hot, rub it with a piece of fresh suet to prevent the meat from sticking, and lay on the steaks, which should be cut an equal thickness about ¾ inch, or rather thinner, and level them by beating them (as little as possible) with a rolling-pin; turn them frequently with steak tongs (if these are not at hand stick a fork in the edge of the fat so that no gravy escapes), and in 8 or 10 minutes the steak will be done. Have ready a very hot dish, into which put the ketchup, and, when liked, a little minced shalot, dish up the steaks, rub them over with butter and season with pepper and salt. They should not be cooked before the time required, as their excellence depends upon being served very hot.

Fried Beef Steak.

Although broiling is a superior method of cooking steaks to frying them, yet, when the cook is not expert, the latter mode may be adopted, and when properly done, the dish may look inviting, and the flavour be good. The steaks should be cut rather thinner than for broiling, and with a small quantity of fat for each. Put some dripping into a frying pan, and let it get quite hot, then lay in the steaks. Turn them frequently until done, which will be in about 8 minutes or rather more should the steaks be very thick ; serve on a very hot dish, season with pepper and salt. They should be sent to table quickly, as, when cold, the steaks are entirely spoiled.

MUTTON CHOPS, BROILED.

Cut the chops from the loin of mutton, remove a portion of the fat, and trim into a nice shape, slightly beat and level them, place the gridiron over a bright, clear fire, rub the bars with a little fat, and place on the chops ; while broiling, frequently turn them, and in about 8 minutes they will be done ; season with pepper and salt, and dish them on a very hot dish, rub a piece of butter on each chop, and serve hot.

EGGS.

Poached.—Break some new-laid eggs into separate cups ; then drop them one after the other into a stewpan containing boiling water, mixed with a tablespoonful of white vinegar and a little salt ; keep this boiling while the eggs are dropped in at the side of the stewpan ; when they have boiled for 2 minutes, drain them on a clean cloth, then place each one on a square or oval piece of dry toast or fried ham, bacon, &c.

Boiled, for breakfast, should be placed in boiling water, and allowed from 3 to $3\frac{1}{4}$ minutes to set the whites nicely; if liked hard, 6 to 7 minutes will not be found too long; for salad, they should be boiled 10 to 15 minutes.

FRIED BACON.

Cut the bacon into thin slices, trim away the rusty part, and cut off the rind; put into a cold frying pan, that is to say, do not place the pan on the fire before the bacon is in it; turn it 2 or 3 times, and dish it on a very hot dish, poach the eggs, and slip them on to the bacon without breaking the yolk, and serve quickly.

OMELETTE WITH FINE HERBS.

Break 3 eggs in a basin; to these add a tablespoonful of cream, a small pat of butter broken into small pieces: a little chopped parsley and shalot, some pepper and salt ; then put 2 ozs. fresh butter in an omelette pan on the stove fire; while the butter is melting, whip the eggs, etc., well together until they become frothy; as soon as the butter begins to fritter, pour the eggs into the pan, and stir the omelette; as the eggs appear to set and become firm, roll the omelette into the form of an oval cushion; allow it to acquire a golden colour on one side over the fire; then turn it out on a dish, pour a little thin sauce round it, and serve.

Barley Soup.

Stock, barley, mixed vegetables, onions, flour, pepper, salt, celery seed.

Scald the barley by pouring boiling water over it, allow to stand for a few minutes, then throw the water away. If this is omitted, the soup will be of a bluish colour.

Proceed as for pea soup, the barley being added after the stock has arrived at the boiling point. Omit the dried mint, celery seed being added with the pulped vegetable.

Pea Soup.

Stock, split peas, mixed vegetables, onions, flour, pepper, salt, dried mint. Place the stock in the boiler, scrape and clean the vegetables and onions, place them in the stock with the peas; as soon as the latter comes to the boil, let it simmer gently until the vegetables are tender, then take them out and pulp them; mix the flour, pepper, and salt into a smooth batter, add the pulped vegetables, and mix well together; bring the contents of the boiler to a sharp boil, add the thickening, and stir quietly until it comes to the boil again, then let it simmer for 30 minutes, and serve; rub the dried mint to a powder, put a little in each can, and pour the soup over it.

Lentil Soup.

Ingredients.—Lentils, mixed vegetables, onions, flour, celery seed, pepper, salt, stock.

Proceed as for pea soup, using lentils instead of peas, adding the celery seed with the thickening.

Lentil and Pea Soup.

Stock, lentils, split peas, mixed vegetables, onions, flour, pepper, salt, mixed herbs 1 packet.

Proceed as for Lentil Soup.

Hotch Potch.

Soak the blue peas over night. Peel or scrape, clean, and cut the vegetables and onions into small square pieces. Scald the barley by pouring boiling water over it, allow it to stand for a few minutes, then throw the water away. Clean and cut the lettuce, or cabbage, into small pieces. Place stock into a boiler, add the peas, barley, vegetables, onions, parsley, cabbage, and herbs. Bring the whole to a simmer, and allow to simmer gently till the peas, barley, &c., are cooked. Make the thickening by mixing the flour, pepper, and salt with a little cold water. Bring the soup to a sharp boil, add the thickening, and keep it stirred till it comes to the boil again, then add the celery seed; allow it to simmer about half-an-hour. Draw the fire and damp the boiler down till time to serve.

Tomato Soup.

Ingredients for 60 men.—Onions, 4 lbs.; tomatoes, 8 lbs. (or 4 tins of preserved tomatoes); harricot beans, 6 lbs.; flour, 1 lb.; salt, 2 ozs.: pepper, 1 oz.; 6 gallons of stock.

Soak the beans for 12 hours in plenty of cold water. Cook them by steam for 3 hours or until they are soft. Meanwhile peel and cut up the onions very small; add them to the stock, also the tomatoes. Allow the contents of the boiler to simmer gently until the onions are cooked and the tomatoes are reduced to pulp. Remove the beans, and pulp them. This can be done either with the usual vegetable masher or by passing them through a mincing machine. Add the flour, pepper and salt and the pulped beans; bring the soup to the boil. Stir in the beans. Allow to cook slowly for 45 minutes, then serve.

Plain Pudding.

Flour, dripping, baking powder, salt. Mix the flour, salt, and baking powder well together, shred the dripping if hard, if soft break it into small pieces; the dripping must not be rubbed into the flour. Add the dripping to the flour, mix together, moisten with sufficient cold water, mixing lightly till it forms a tolerably soft dough. Divide into equal portions, roll and tie each securely in a cloth, which has been previously wrung out of boiling water and floured, leaving room for the pudding to swell. Boil for 3 to 4 hours.

Plum Pudding.

Flour, raisins, currants, dripping, salt, baking powder, treacle, or sugar. Stone and chop up the raisins, or use sultanas; wash, dry, and carefully pick the currants, add them to the flour, dripping, salt, and baking powder; mix the treacle, or sugar with the water required, and add to the remaider and stir well together.

Tie up in a cloth as in plain pudding, and boil gently for 3 hours; spice, sugar, or eggs added is an improvement. This pudding may also be cooked by steam. Have the required number of basins (one being sufficient for two men), rinse them in water, and equally divide the pudding; then proceed as for meat puddings.

Jam Rolls.

Ingredients.—The same as for paste, with the addition of jam. Prepare the paste, and roll it out until it is about ½ inch thick, spread the jam over it, damp the edges of the paste and roll it up, care being taken that the paste at the ends adheres to each other to prevent the jam boiling out, roll up in a cloth as with a plain pudding, and boil gently for 3 hours.

CURRANT ROLLS.

Ingredients.—The same as for plain pudding, with the addition of currants and sugar.

Proceed as for jam rolls, wash and dry the currants, picking out the stalks and any grit that may remain, distribute the currants equally over the paste; add a thin layer of sugar, roll and finish as for jam rolls.

RAISIN PUDDING.

Flour, raisins, dripping, egg powder, baking powder, salt.

Stone and chop up the raisins if time will allow; if not, use sultana raisins, shred the dripping, then mix the flour, baking powder, egg powder, salt, and dripping, add the raisins, and mix well together; add sufficient water to make a rather stiff paste, divide it into equal portions, tie in a cloth, and boil for 4 hours; if rolled, as in a plain suet pudding, 3½ hours will be found sufficient.

DATE PUDDING.

Dates, flour, sugar, dripping, salt, nutmegs.

Stone the dates, shred the dripping, place the flour, salt, sugar, and grated nutmeg in a dish, mix together, add the dates and dripping, mix the whole well together; moisten with sufficient cold water (mixing lightly) to make a tolerably soft dough.

Tie up in cloths as in plain pudding, and boil gently for 3½ hours.

BAKED RICE PUDDING.

Rice, milk, sugar, nutmegs, butter.

Wash the rice in cold water and boil until nearly tender, strain the water from the rice, butter the sides of the baking dish, mix the milk and sugar together, divide the rice equally between the dishes, and well mix with the milk and sugar; distribute the butter in small pieces over the surface, grate a little nutmeg over the top, and bake in a moderate oven for 1 hour. Slices of candied peel, currants, or sultana raisins may be added to improve the flavour.

BREAD AND BUTTER PUDDING.

Bread, sugar, currants, butter, milk, suet.

Cut the bread into moderately thin slices (remove any hard crust), butter it; wash, dry, and carefully pick the currants, free the suet from skin and chop fine; place a layer of bread at the bottom of the dish, a layer of currants and sugar, and suet, then another layer of bread, and so on alternately, till the dish is nearly full. Pour in the milk at the side of the dish until it appears on the surface. Bake of a nice brown colour in a moderate oven. It will require 1½ hours.

TAPIOCA.

Tapioca, milk, sugar, nutmegs, butter; soak the tapioca in a small quantity of water, divide it equally in the dishes, add the sugar to the milk, and well mix with the tapioca, break the butter in small pieces, distribute over the surface of the pudding, grate a nutmeg over each dish, and bake for 1 hour or longer, according to the size of the pudding.

APPLE PIES OR TARTS.

Flour, apples, dripping, baking powder, sugar, cloves.

Make the paste described in No. 1, peel, core, and cut the apples into slices, place a thin border of paste round the sides of the dishes, and add the sugar and cloves, with sufficient water to cover the bottom of the dish, cover with paste, and bake in a quick oven for 1 hour.

APPLE PUDDING.

Flour, apples, baking powder, sugar, cloves, salt, dripping.

Make the paste, peel, core, and cut the apples into slices. Line the inside of a "Dean's" or "Warren's Cooker" with a portion of the paste. Place in the apples, sugar, and cloves, well cover the bottom of the cooker with water, cover with the remainder of the paste, and with a moderate steam will require from 2 to 2½ hours to cook.

APPLE RINGS.

Apple rings, or any fruit from which the moisture has been evaporated, should be soaked in the same quantity of water for eight or nine hours. They are then used as detailed for apple tarts, care being taken, however, to see that they do not get dry during the process of cooking. Dried or evaporated fruits require more time to cook and absorb more water than fresh fruit.

DRIED FIGS AND PRUNES.

If for stewing, they should be separated and picked over, then soaked over night in sufficient water to cover them. Figs or prunes may be stewed by means of steam, or in a covered dish in an oven. Sufficient sugar to sweeten them should be added, and if required, a little lemon flavouring.

Figs for puddings should be prepared as for stewing, the hard stems removed and the fruit cut into small pieces, then proceed as for Date Pudding.

TREACLE PUDDING.

Flour, treacle, baking powder, egg powder, salt, dripping.

Place the flour, baking and egg powders, with salt in a dish, mix well. Put in the chopped dripping, add sufficient cold water to make a tolerably stiff paste.

Roll out about half an inch thick, spread the treacle over the surface of the paste, damp the edges, roll round, taking care that the ends adhere to each other to prevent the treacle from boiling out. Tie in a cloth, and boil gently from 2½ to 3 hours.

TREACLE TARTS.

Flour, treacle, dripping, baking powder, and bread crumbs.

Make the paste. Grease the inside of a pie dish. Roll out the paste to about one-eighth of an inch thick; line the inside of the dish; spread a thin layer of treacle at the bottom; sprinkle on the bread crumbs. Cut a piece of paste the size of the dish. Place this on the top, add another layer of treacle, then a layer of paste, and so on till the dish is nearly full; bake in a moderate oven till done; time required for baking depends upon the size of the dish; an ordinary dish about 1½ hours.

MACARONI PUDDING.

Ingredients for 60 men.—6 lbs. macaroni; 3 gallons milk; 3 lbs. sugar; 1 nutmeg; ¼ lb. butter.

Break the macaroni into pieces about 1 inch long. Drop them into boiling water to which a little salt has been added, and simmer for about 40 minutes.

Proceed as for Rice Pudding.

BISCUIT PUDDING.

Ingredients for 60 men.—10 lbs. biscuit crumbs; 4 lbs. flour; 3 lbs. sugar; 3 lbs. suet; 41 tablespoonfuls lemon or lime juice. Pinch of salt.

Method.—Thoroughly dry the biscuits, then crush with a rolling pin or pass through a mincing machine. Soak in cold water for about 15 minutes. Remove all skin from the suet, then chop very fine. Add the flour, sugar, suet, salt and lime juice to the soaked biscuits. Stir well together with sufficient water to make a stiff paste. Grease some basins or dishes and boil or bake till done. The time allowed for cooking depends on the size of the pudding.

BREAD CRUMB PUDDING.

Ingredients for 60 men.—8 lbs. bread crumbs; 4 lbs. flour; 3 lbs. currants; 3 lbs. raisins; 3 lbs. suet; 3 lbs. sugar; a little spice and a pinch of salt.

Method.—Carefully prepare the fruit, then proceed as for biscuit pudding.

BAKED CUSTARD PUDDINGS FOR 5 OR 6 PERSONS.

1½ pints of milk, the rind of ¼ lemon, ¼ lb. of moist sugar, and 4 eggs. Put the milk into a saucepan with the sugar and lemon rind, and let this infuse for half an hour, or until the milk is well flavoured, whisk the eggs, yolks and white; add the milk to them, stirring all the while; then have ready a pie dish, lined at the edges with paste already baked; strain the custard into the dish, grate a little nutmeg over the top, and bake in a very slow oven for about half an hour or a little longer; the flavour of the pudding may be varied by substituting bitter almonds for the lemon rind, and it may be much enriched by using half cream and half milk, and doubling the quantity of eggs.

Arrowroot Blanc Mange for 5 or 6 Persons.

4 large tablespoonfuls of arrowroot, 1½ pints of milk, 3 laurel leaves, or the rind of ½ a lemon, sugar to taste; mix to a smooth batter the arrowroot with ¼ pint of milk; put the other pint on the fire, with laurel leaves or lemon peel, whichever may be preferred, and let the milk simmer until it is well flavoured. Then strain the milk and add it boiling to the mixed arrowroot; sweeten with sifted sugar, and let it boil, stirring it all the time, till it thickens sufficiently to come from the saucepan; grease a mould with pure salad oil, pour in the blanc mange, and when quite set, turn it out on a dish, and pour round it a compote of any kind of fruit, or garnish it with jam. A tablespoonful of brandy, stirred in just before the blanc mange is moulded, very much improves the flavour of this dish. Cost 6*d*. with the garnishing.

Cheap Blanc Mange.

¼ lb. sugar, 1 quart milk, 1½ ozs. isinglass, the rind of ½ lemon, 4 laurel leaves. Put all the ingredients into a lined saucepan, and boil gently until the isinglass is dissolved; taste it occasionally to ascertain whether it is sufficiently flavoured with the laurel leaves; then take them out and keep stirring the mixture over the fire for about 10 minutes; strain it through a fine sieve into a jug, and when nearly cold, pour it into a well oiled mould, omitting the sediment at the bottom, turn it out carefully on a dish, and garnish with preserves, bright jelly, or a compote of fruit.

Baked Rice Pudding for 5 or 6 Persons.

Small teacupful of rice, 4 eggs, 1 pint milk, 2 ozs. fresh butter, 2 ozs. beef marrow, ¼ lb. of currants, 2 tablespoonsfuls of brandy, 1 nutmeg, ¼ lb. sugar, and the rind of ½ a lemon.

Put the lemon rind and milk into a stewpan, and let it infuse until the milk is well flavoured with the lemon; in the meantime, boil the rice until tender in water, with a very small quantity of salt, and when done, let it be thoroughly drained; beat the eggs, stir in them the milk, which should be strained; the butter, marrow, currants, and remaining ingredients; add the rice and mix the whole together, line the edge of the dish with puff paste, put in the pudding and bake in a slow oven for about three-quarters of an hour. Slices of lemon peel may be added, or sultana raisins may be substituted for the currants.

Tapioca Pudding.

Put 10 ozs. of tapioca into a stewpan with a quart of milk, 6 ozs. of sugar, 2 ozs. butter, a pinch of salt, and some grated lemon peel, stir this over the fire till it boils, then withdraw it; add 4 eggs, mix well and bake for half-an-hour in a pie dish. If the eggs be whipped separately, and gently mixed in with preparation, the pudding will be much lighter. All kinds of farinaceous substances may be prepared as above.

Plain Bread Pudding for 5 or 6 Persons.

Odd pieces of crumb of bread, salt, grated nutmeg, moist sugar, currants, and butter. Break the bread into small pieces, and pour as much boiling water on them as will soak them well; let these stand until the water is cool, then press it out and mash the bread with a fork until it is quite free from lumps.

Measure this pulp, and to each quart add ½ teaspoonful of salt, 1 of grated nutmeg, 3 ozs. moist sugar, and ¼ lb. currants; mix it all well together, and put it into a well-buttered pie dish; smooth the surface with the back of a spoon, and place a small piece of butter on the top; bake in a quick oven for 1½ hours, and serve very hot. Boiling milk substituted for the boiling water would very much improve the pudding. Cost 6d.

Bread and Butter Pudding.

2 lbs. bread, ¼ lb. butter, 1½ pints of milk, 4 eggs, sugar to taste, ¼ lb. currants, flavouring of vanilla, grated lemon peel or nutmeg.

Cut the bread into slices, and butter them and place in a pie dish, with currants between each layer and on the top, sweeten and flavour the milk, either by infusing a little lemon peel in it, or by adding a few drops of essence of vanilla; well whisk the eggs and stir these to the milk, strain this over the bread and butter, and bake in a moderate oven for 1 hour or more. This pudding may be very much enriched by adding candied peel or more eggs than stated above. It should not be turned out, but sent to the table in the dish, and is better if made about 2 hours before being baked.

Boiled Rice Pudding.

¼ lb rice, 1½ pints milk, 2 ozs. butter, 4 eggs, ½ oz. salt, 4 large tablespoonfuls of sugar, flavouring to taste. Stew the rice very gently in the milk, and when it is tender pour it into a basin; stir in the butter and let it stand to cool, then beat the eggs; add these to the rice with the sugar, salt and any flavouring that may be approved, such as nutmeg, powdered cinnamon, grated lemon peel, essence of bitter almonds, or vanilla; when all is well stirred, put the pudding into a buttered basin; tie it down with a cloth; plunge it into boiling water, and boil for 1¼ hours.

BREAD PUDDING.

Bread, currants or raisins, candied peel, sugar, chopped suet, flour, milk, baking powder, and salt.

Soak the bread in warm water for 15 minutes; squeeze the bread as dry as possible; add the fruit, chopped suet, flour and salt; mix well together; dissolve the baking powder in the milk; add this to the other ingredients, stirring well; a little spice, ginger or grated nutmeg may be added to improve the flavour.

Grease the inside of a dish; place in the mixture, smoothing the surface with the back of a tablespoon; steam or boil for 3 hours.

PLAIN PANCAKES.

Mix in a basin with a spoon 4 ozs. of flour, 4 eggs, a little salt, some grated lemon peel, and a pint of milk or cream, and fry spoonfuls of this batter with a little butter in small frying pans over a clear fire.

The pancakes must be fried on both sides, and when done rolled up with sugar inside, and dished up on a warm dish. French pancakes are made by introducing some preserve in the ordinary pancake.

PANCAKES (SUGAR).

Put the pan on the fire with a tablespoonful of dripping, let it melt; pour off all that is not wanted, then pour in 3 tablespoonfuls of the following batter:—

Break 4 eggs in a basin, add 4 small tablespoonfuls of flour, 2 teaspoonfuls of sugar, a little salt, beat all well together, mixing by degrees half a pint of milk a little more or less depending on the size of the eggs and the quality of the flour; it must form a rather thick batter, a little ginger, cinnamon, or any other flavour can be added if preferred, 2 eggs only may be used, but in this case use a little more flour and milk. When set and one side brownish, lay hold of the pan at the extremity of the handle, give it a sudden but slight jerk upward, and the cake will turn over on the other side, which when brown, dish up with sifted sugar over; serve with lemon; chopped apples may be added to the batter; currants and sultanas can be mixed with it.

MUFFINS.

Ingredients.—Flour, eggs, milk, butter, carbonate of soda, tartaric acid or baking powder, and salt.

Place 2 lbs. of flour in a dish; add a good pinch of carbonate of soda, and tartaric acid and a little salt, mix the whole well together; melt about two oz. of butter; add it to the flour; mix lightly; add the eggs and milk, which have been previously well whisked together; stir lightly until it becomes a nice light paste; take it out of the dish and roll it out about three-quarters of an inch thick; care being taken that the ingredients are handled as lightly as possible in mixing and rolling; cut the muffins out a round or triangular shape; place in a hot oven for a few minutes, care being taken that they are turned; 1 egg and ¼ pint of milk will be found sufficient for the quantity. In using baking and egg powder, ½ teaspoonful of each will be sufficient.

TEA SCONES.

Ingredients.—Flour, eggs, milk, baking powder, salt and butter. Proceed as above, roll them out a little thinner, and bake them in a hot oven or on a griddle a nice brown colour, and serve hot. (This remark applies to muffins.)

TEA SCONES.

Ingredients the same as above, substituting dripping for butter. Get some nice beef dripping, place the flour and baking powder in a dish, rub the dripping in the flour, and mix the whole well together, add the milk and eggs as before, mix lightly, and proceed as above.

SODA SCONES.

3½ lbs. of flour, large teaspoonful of carbonate of soda, 1 teaspoonful of cream of tartar, buttermilk, and a small teaspoonful of salt.

Mix the dry ingredients well together, lightly add the butter and milk to make the dough, and divide into from 4 to 6 pieces. Sprinkle a little flour on the baking board, and roll out the dough with the rolling pin to about a ¼ of an inch thick, cut in four, and bake on a hot griddle till of a pale brown colour, then turn and bake the other side.

WHEATEN MEAL SCONES.

1 lb. wheatmeal, 1 lb. flour, teaspoonful of carbonate of soda, teaspoonful of cream of tartar, teaspoonful of dripping, half a teaspoonful of salt, and a little buttermilk.

Mix the meal, flour, soda, cream of tartar, dripping, and salt well together, then add the buttermilk to make a light dough, divide, and roll out to the thickness of ¼ of an inch, and bake on not too hot a griddle.

RICE SCONES.

1 lb. rice, ¼ lb. flour, 1 teaspoonful of sugar, and ⅓ teaspoonful of salt. Put the rice and sugar into a saucepan with 1 quart of water, and let it come to the boil; then set it to the side of the fire, and let it steam for 2 hours with the lid closed till all the water has been absorbed, and the rice becomes soft; then sprinkle the flour on the baking board, and turn the rice on it, let it stand to cool, then divide into 6 parts and roll out very thin, cut each in 3, and bake on not too hot a griddle.

BEEF TEA.

To each pound of beef allow 1 quart of water. Pare away every particle of fat and cut the meat (which should be cut from the rump or gravy piece) into very small squares of mince, and put this into a clean stewpan, add the water and set it on the fire to boil, remembering that as soon as the scum rises to the surface it should be removed with a spoon, and a very small quantity of cold water and salt should be poured in at the edge of the stewpan in order to facilitate the rising of the albumen in the form of scum. Unless due precaution be taken to effect the skimming satisfactorily the broth, instead of becoming clear and bright, becomes thick and *turbid*, and thus presents an unappetising appearance to the eye of the patient.

When beef tea has boiled gently for about half an hour and become reduced to about half its original quantity, let it be strained through a clear sieve or napkin into a basin, and serve with dry toast and salt. The foregoing is intended for patients whose case may require comparatively weak food; in cases where food of a more stimulating character is needed in the form of extract of beef, it will be necessary to double the quantity of meat, and when it happens that beef tea is required in a hurry the meat should be chopped as finely as sausage meat, put into a stewpan with boiling water, stirred on the fire for ten minutes, and then strained through a napkin for use.

MUTTON BROTH.

To each pound and a half of the scrag of mutton add 1 quart of water, a little salt, 2 ozs. of pearl barley. Chop the mutton into small pieces and add with the water in the stewpan; set it to boil, skim it well, add a little salt and the barley, boil gently for 1 hour, strain off the broth through the sieve into a basin, and serve with dry toast; a turnip and half a head of celery may be added where vegetables are not objected to.

CHICKEN BROTH

Draw the chicken, scald the legs, and remove the cuticle which covers them, cut up the chicken into members or joints, leaving the breast whole, put the pieces of chicken into a very clean stewpan, with a quart of water, a little salt, and 2 ozs. of washed rice, boil very gently for 1 hour, and when done serve the broth with or without the rice, according to taste.

RICE WATER.

Wash 3 ozs. of rice in several waters and then put in a clean stewpan with a quart of water and 1 oz. of raisins, boil gently for $\frac{1}{2}$ an hour, strain through a coarse hair sieve into a jug.

BARLEY WATER.

2 ozs. of pearl barley boiled in a quart of water for 20 minutes and afterwards allowed to stand until it becomes cold; it must then be strained through a sieve into a jug, and a small piece of lemon peel added.

TOAST AND WATER.

Boil a quart of water and pour it on a good-sized piece of crumb of bread which has been well toasted before a clear fire until it becomes nearly crisp and of a dark brown colour; allow this to steep for half an hour; it is then ready.

SUGAR WATER.

To a pint of cold spring water add an ounce of lump sugar and a tablespoonful of orange flower water, mix. This is a very refreshing drink in summer, and is besides perfectly harmless.

ARROWROOT.

To half a pint of boiled water add rather more than half an ounce of Bermuda arrowroot, previously mixed in a teacup with a wineglassful of cold water. Stir this on the fire until it boils for a few minutes, pour it into a basin, flavour with a little sugar, and a small spoonful of brandy or a little red or white wine, or else with a little piece of orange or lemon peel, which may be boiled with the arrowroot.

TO PREPARE SAGO OR TAPIOCA.

Boil 2 ozs. of either in a pint of water for 20 minutes, and flavour as directed for arrowroot; sago may also be boiled in either mutton, chicken, or veal broth, or in beef tea.

TO MAKE GRUEL.

Take one teaspoonful of oatmeal and mix with a wineglassful of water, and having poured this into a stewpan containing a pint of boiling water, stir the gruel on the fire, to boil ten minutes; pour it into a basin, add salt and butter, or if more agreeable, rum, brandy, or wine and sugar.

OATMEAL PORRIDGE.

Boil a quart of water in a saucepan, as soon as it boils sprinkle slowly in a cupful of coarse oatmeal, stirring gently until it is thick and smooth enough, pour it at once on to plates and serve with cold milk or treacle.

ONION PORRIDGE.

Take a large Spanish onion, peel and split it into quarters, and put these into a small saucepan with a pint of water, a pat of butter and a little salt, boil gently for half an hour, add a pinch of pepper, and eat the porridge just before retiring for night. This is an excellent remedy for colds.

TO MAKE STOCK FOR JELLY AND CLARIFY IT.

Ingredients.—2 calves' feet, 6 pints water. The stock for jellies should always be made the day before it is required for use, as the liquor has time to cool, and the fat can be so much more easily and effectually removed when thoroughly set. Procure 2 calves' feet, scald them to take off the hair, slit them in two, remove the fat from between the claws, and wash the feet well in warm water, put them into a stewpan, with the above proportion of cold water, bring it gradually to the boil, and remove every particle of scum as it rises; when it is well skimmed boil very gently for 6 or 7 hours, or until the liquor is reduced rather more than one half; then strain it through a sieve into a basin, and put it into a cool place to set; as the liquor is strained, measure it to ascertain the proportion for the jelly, allowing for the sediment and fat at the top. To clarify it, carefully remove all the fat from the top, pour over a little warm water to wash away any that may remain, and wipe the jelly with a clean cloth; remove the jelly from the sediment, put it into a saucepan, and, supposing the quantity to be a quart, add to it 6 ozs. of loaf sugar, the shells and well-whisked whites of 5 eggs, and stir these ingredients together cold; set the stewpan on the fire, but do not stir the jelly after it becomes warm; let it boil about 10 minutes after it rises to a head, then throw in a teacupful of cold water, let it boil for five minutes longer, then take the saucepan off, cover it closely, and let it remain ½ an hour near the fire; dip the jelly bag into hot water, wring it out quite dry, and fasten it on to a stand or the back of a chair, which must be placed near the fire to prevent the jelly setting before it has run through the bag; place a basin underneath to receive the jelly; then pour it into the bag, and should it not be clear the first time, run it through the bag again. This stock is the foundation of all really good jellies, which may be varied in innumerable ways by colouring and flavouring with liquors, and by moulding it with fresh and preserved fruits. To ensure the jelly being firm when turned out, ½ oz. isinglass, clarified, may be added to the above proportion of stock. Substitutes for calves' feet are now used plentifully in making jellies, which lessen the expense and trouble in preparing this favourite dish, isinglass and gelatine being two of the principal materials, but although they may look as nicely as jellies made from good stock, they are never so delicate, having very often an unpleasant flavour, somewhat resembling glue, particularly when made with gelatine.

Cowheel Stock for Jellies.

Procure 2 heels that have only been scalded and not boiled, split them in two, and remove the fat from between the claws; wash them well in warm water, and put them into a saucepan with 3 quarts of cold water, bring it gradually to the boil, remove all scum as it rises, and simmer the heels gently for 7 or 8 hours, or until the liquor is reduced one-half, then strain it into a basin, measuring the quantity, and put it into a cool place; clarify it as directed for calves' feet, using with the other ingredients about ½ oz. isinglass to each quart. This stock should be made the day before it is required for use. Two dozen shank bones of mutton boiled for 6 or 7 hours yield a quart of strong, firm stock. They should be put on in 2 quarts of water, which should be reduced one-half. This should also be made the day before it is required.

Bread Making.

Bread may be broadly divided into two classes:—
(a) Fermented, or leavened bread, in which the carbonic acid gas necessary to distend the dough and cause the loaf to rise is produced by some form of yeast.
(b) Unfermented bread, in which the requisite gas is either produced by chemicals (baking powder), or forced into the dough by a mechanical process.

Fermented bread is usually made in the army, though baking powder may be resorted to on active service, or on extreme emergency.

Fermented bread is manufactured from wheat flour, water, salt, and some form of yeast.

It is necessary to allude to these ingredients briefly, before describing the process of bread making.

Flour.

The conditions of contract enact that flour shall be the produce of good, sound, sweet and dry wheat, without any adulteration whatever, and of such a grade that a sack of 280 lbs. will produce at least 180 2-lb. loaves of bread of the " Best Household " quality. This last clause refers to what is known as the " gain per cent.," by which is meant the difference in weight between 100 lbs. flour and the weight of bread produced from that amount of flour. Good flour absorbs half its own weight of water, but a large proportion of this extra moisture is lost by evaporation during baking; and experience has shown that a nett gain should remain of not less than 80 per cent., as laid down in the conditions of contract.

Water.

Soft water is best, and it is essential that it should be clean and pure.

Salt.

The functions of salt are to bind the dough, to prevent injurious fermentation, and to impart a flavour to the loaf. It should be white, crystalline, dry, and soluble in water. The usual proportion is 3½ lbs. to every sack of flour (280 lbs.).

Yeast.

Yeast is a plant of the fungus tribe, which in congenial soil grows very rapidly, and gives off large quantities of carbonic acid gas.

This gas is employed in raising the bread, and making it light and digestible.

"*Parisian Yeast*" is the yeast generally used in the Service in the field, and is made by the baker himself on the spot as follows:—

To make 1 gallon, boil 1 gallon of water, put into it 1 oz. hops, and allow them to simmer 40 minutes. Take ½ lb. flour, mix it with a little cold water, and scald it with sufficient of the hop liquor to make a thick paste. Then strain the remainder of the hop liquor on to the paste, thoroughly mix, and allow the mixture to cool down to 90°. Then "stock," *i.e.*, introduce the yeast plant, using any substance containing that plant in large quantities, such as 1 pint of old "Parisian" yeast, ½ oz. D.C.L. yeast, 1 pint of beer or stout, or ½ lb. sugar. The yeast should be allowed to rise and fall once before being used. 5 pints are required to each sack of flour (280 lbs.).

Sour Dough Yeast is also manufactured by the baker, and may be used on emergency when "Parisian" or other yeast is not available. It is made as follows:—

Mix about 4 lbs. of flour with water into a dough, and allow the mass to ferment for about 12 hours in a warm atmosphere. Directly the mixture shows the slightest signs of movement, add 2¼ gallons of water with 1 oz. salt dissolved in it. This preparation is quick-working, but unreliable and difficult to handle. 6 lbs. of dough in 2¼ gallons of water are required to each sack of flour.

The various operations which take place in turning the above ingredients into bread are as follows:—

1. Setting the sponge.
2. Making the dough.
3. Scaling and moulding.
4. Baking.

It is assumed throughout that two sacks of flour (560 lbs.) are being converted into bread, this being the most convenient quantity for bakers to handle.

Setting the sponge.

The sponge is a preliminary mixture of part of the flour with the total amount of yeast necessary, and a due proportion of water, and its object is to give the yeast a fair chance to get firmly established.

Sponges are described as "¼ sponge," "½ sponge," and "¾ sponge," the fraction indicating the proportion of the total amount of flour to be used in setting the sponge. The class of sponge to be used varies according to conditions of climate and temperature, *i.e.*, in a frost the ¼ sponge might be used, whereas in a very hot climate the ¾ sponge would be employed. In this country the ½ sponge is the one most commonly worked with.

Assuming that a ½ sponge is used with Parisian Yeast, the component parts of it, *i.e.*, 280 lbs. flour, 10 pints of yeast and 12¾ gallons of *warm* water, are thoroughly mixed together in a trough, and allowed to rise and fall twice. This takes about 12 hours, when the mixture is ready for the next process, which is:

Making the Dough.

The sponge is now broken up, and the remaining ingredients, *i.e.*, 280 lbs. flour, 7 lbs. salt, and 14 gallons of water are added, and the whole mass thoroughly kneaded. It is then left to work for about 3 hours.

Scaling and Moulding.

The dough is now turned out of the trough, cut into lumps, and the lumps "scaled" according to the weight of the loaves required. As considerable weight is lost by evaporation during baking, and up to the time of issue, a 2 lb. loaf must be scaled at 2 lbs. 3½ ozs., and a 2½ lb. loaf at 2 lbs. 13 ozs.

The scaled lumps are then moulded or shaped into the form of a loaf, left in a warm place for about ¼-hour, and are then ready for the oven.

Baking.

The moulded lumps of dough are then placed into the oven.

The proper heat for an oven is from 400° to 500°, according to the class of oven used, and the time taken in baking is as follows:—

	In Tins or separate Loaves.	In Batch Bread.
2½ lb. loaf	50 min. to 1 hr.	1¼ hrs.
2 lb. „	40 „ to 1 hr.	1¼ hrs.

When baked, bread should be taken at once into the bread store to cool. The store should be dry, cool, and well ventilated, and not more than two layers of loaves of new bread should be placed on one rack.

Bread made with Baking Powder.

The advantage of using baking powder is the saving of time effected in the production of bread, a feature which may sometimes render this process useful on service when bread has to be produced at short notice. The method of using baking powder is as follows:—

Spread the flour evenly at the bottom of the trough, sift the baking powder over the flour, taking care to break up any small lumps, which, if left, would cause a yellow stain in the bread. The dry powder and flour should then be thoroughly mixed.

Dissolve salt, at the rate of 2½ lbs. per sack of flour only, as a considerable quantity of saline matter is contained in all baking powders, in the softest and coldest water obtainable; water which has been boiled and allowed to get cool is the best for the purpose.

Mix the flour, baking powder, and water thoroughly with a rotary motion, constantly stirring up from the bottom. The dough being properly mixed should be scaled, moulded, and placed in a quick oven. To make a good loaf with baking powder, the bread should be in the oven within 30 minutes of adding the water to the flour. No more salt than the above-mentioned proportion should be used, or the bread becomes heavy, dark, and briny. If the dough is allowed to lie about, the effervescence is finished before it is put in the oven. Full directions as to the method of using baking powders, and the proportion required, are given on the tins.

Judging Bread.

The current conditions of contract, a copy of which should be hung up in every bread store, enact that the bread supplied shall be sweet, well made, properly baked, and of the description or quality known as "best household," made from flour clean and free from grit, the produce of good, sound, sweet, and dry wheat; that it shall be in all respects as good in quality as the best plain or fine (as distinguished from fancy) bread usually sold by the trade as "Best Household Bread," with which it shall frequently be compared.

The bread must be delivered not earlier than 24 hours, nor later than 48 hours after baking, and the loaves must weigh 2 lbs. *at the time of issue*.

It is subject to inspection and approval by an Officer or Officers acting on behalf of the G.O.C., and in case of rejection the contractor has the right of appeal to the Officer Commanding at the station, and finally to the General Officer Commanding.

The main characteristics of a good loaf, fulfilling the conditions enumerated above, are as follows:—

The crust should be a rich yellowish-brown, well baked but not burnt, as thin as possible, and distributed all round the loaf. The crumb should be cream-white in colour, light, flaky, elastic, and full of small, evenly distributed cavities.

In tasting a loaf, the crumb should always be eaten.

Several loaves should be selected from different parts of a consignment of bread, and each weighed singly.

When required, the contractor must deliver bread in accordance with the specification, to the extent of half a pound per diem for each soldier included in the ration return for use in the Regimental Recreation Rooms, and for other similar purposes. The right is also reserved to issue Biscuit from Government Stores to the extent of one issue per week.

VARIETIES OF BREAD.

In England, as a rule, bread is exclusively made of wheat flour. Abroad, however, the flour of other cereal grains is also used. Wheat flour is by far the most suitable for bread baking, being the most nutritious, and containing a larger proportion of gluten than other flour.

Indian corn bread, although less nutritious than that made from wheat, is more fattening, in consequence of the greater quantity of oil it contains. It does not bake in the light spongy loaves as wheaten flour, and its flavour is not agreeable. It is, however, excellent in the form of cakes.

Rye bread is little used in England, but common on the Continent. It is wholesome but dark coloured, sometimes black and less spongy than bread made from wheat flour. It possesses the quality of retaining its freshness for a long time.

Oat bread. Owing to a peculiar quality of the gluten which the oat contains, the meal of this grain does not admit of being baked into a light spongy bread.

Rice bread. Rice flour is scarcely ever made into bread, although it is not infrequently mixed with wheat flour intended for bread, and sold under the name of "corn flour." It is cheaper than wheat flour, and is used for dusting the boards, troughs, and dough.

FIELD INSTRUCTION.

To cook rapidly and well is an art which can be easily acquired, and which every soldier should learn. Officers commanding are responsible that there are a certain number of men (at least 8 or 10) in each troop or company who have been instructed in the cutting up of meat, making field kitchens, and cooking. The serjeant-cook is specially trained for the purpose of instructing men in this part of their duty. It is a matter of paramount necessity that soldiers' food should be carefully looked after, and this should be attended to by the officers themselves.

Service kettles are as follows:—

Name	Weight.	Contents.	Surface Diameter.	Depth outside measure.	Number of men will Cook for. With vegetables.	Without.
	Lbs	Galls.	In. In.	In.		
Oval, large	8	3	13½ by 9	11	8	15
,, small	4¾	1¾	12½ by 8½	8	5	8

On arrival in camp the cooking party, consisting of the serjeant-cook, assistant-cook, two men per troop or company, will proceed to make the kitchen. If the encampment be only for one night, one trench per company should be dug 6 feet long, 9 inches wide, and 18 inches deep at the mouth, and continued for 18 inches up the trench, then sloping upwards to 4 inches at the back with a splay mouth pointing towards the wind 2 feet 6 inches by 3 feet 8 inches deep, and a rough chimney 2 feet high at the opposite end, and formed with the sods cut off the top of the trench. It will be advantageous if these trenches be cut on a gentle slope.

All brushwood and long grass should be carefully cut for a circle of 20 feet round the kitchen, and may be used in lighting the fire. On damp or marshy sites a wall trench will be found to answer best, constructed as follows:—

Cut some sods of turf about 18 inches long and 9 inches wide and lay them in two parallel lines 6 feet long, with an interval between them of 2 feet 6 inches. Build these walls 2 feet high for the large kettles, and 18 inches for the small ones. Lay the wood all over the bottom between the walls Light the fire. This trench will hold 12 large or 20 small ones. It should be built sideways to the wind to prevent the flame and heat being carried through by the draught. If there be no time to dig a trench, or the ground be hard or sandy, the kettles may be placed in rows 10 inches apart and the fires lighted between them, the heat being thus applied to the sides as well at the bottom. If necessary, a row of kettles can be placed across the others over the fire. By this method, however, the cooking takes a little longer and more fuel, but the time required to construct the kitchen is saved. Troops should, in all circumstances, have their dinners an hour and a half after the rations are issued.

Another way is to have a hole prepared by forming a mound with stones, clay, or turf, and making a hollow in the centre the size of the kettle, in such a way as to allow only air enough to support combustion, and prevent the escape of the heat.

The serjeant-cook will apportion the meat, potatoes, &c , to the various messes, which the cooks will cut up and place in the kettles.

Messes should be by kettles—that is, the number of men composing a mess should depend on the kettle used. Lighting the fires should be performed by a man used to the work. Small pieces of wood about the size of lucifer matches should be first ignited and the fire gradually fed with larger ones. By this time the water party should have brought the requisite water in the camp kettles, and the moment the fire is well lighted the kettles should be laid on the trench and brought to a boil, after which allowed to simmer gently. The time from the opening of the ground until the water boils should not exceed 35 minutes; the water in which the potatoes are boiled should not be thrown away, as it is required for washing-up. When the cooking is done for the day, kettles should be filled overnight with clean water and placed on the trenches and covered with turf, so that in the event of rain during the night the trenches and wood may be kept dry. In case the corps move away the cooks should light the fire 30 minutes before reveille, so that the water is boiling by the time it is sounding. The camp kettles should be delivered to the quartermaster by the assistant-cooks of each company. Each cook to fill up the trench he dug. All offal and refuse that is not sold to be buried. All wood not used to be left in a heap on the ground. The serjeant-cook on the line of march should always arrange to have a portion of dry wood carried from one camp to another for kindling purposes. Room can easily be found in the wagons to carry a small bundle of sticks; if not, each cook should carry enough small dry wood to light his own fire. This will be found a great assistance in wet weather.

If troops remain in camp more than a day or two, it is advisable to build a regular field kitchen.

THE ALDERSHOT "GRIDIRON" KITCHEN.

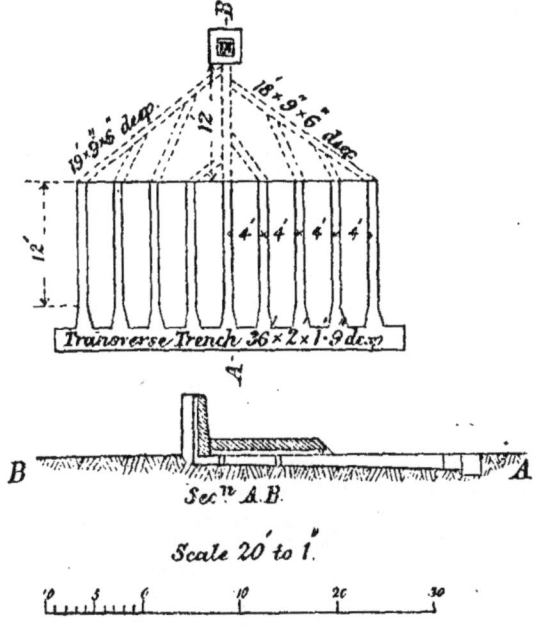

Chimney 6 feet high, 3 feet square at bottom, sloping to 2 feet at top. The trenches are 12 feet long, 9 inches wide, 18 inches deep at mouth, and continuing so for 18 inches in the trench, then sloping to 6 inches on entering the flue.

The gridiron kitchen (Aldershot pattern) consists of 9 trenches 12 feet long, 9 inches wide, 18 inches deep at the mouth, this depth is carried for 18 inches inwards, and forms the fire-place, gradually diminishing to 6 inches where it enters the flue, they are connected by splay mouths 2 feet by 2 feet, and 18 inches deep to the transverse trench, which is 36 feet long, 2 feet wide, and 21 inches deep.

The centre trench is connected with the chimney (6 feet high 3 feet square at the bottom, sloping gradually up to 2 feet square at the top) by a flue 12 feet long, 9 inches wide, and 6 inches deep, covered with the sods removed from the trenches.

To mark out the kitchen, drive a picket to mark the centre of the chimney, a second one 12 feet below, which will mark the top centre of the trench, the third one 12 feet below marking the bottom centre of the trench, the fourth one 2 feet below marks the centre of the splay mouth, and the fifth 2 feet below the outer edge of the transverse trench, for a single trench this would be 4 feet long. For each extra trench added a picket would be driven in 4 feet from each of the latter 4 pickets used in forming the centre or main trench, and parallel to it, leaving after the excavations 3 feet 3 inches for the cooks to work in, the top of each trench being attached to the chimney by a covered flue as shown in diagram.

Where it is possible to build the kitchen on a slope, flues are not required, the trench should be lengthened 1 foot, and a chimney about 2 feet high will be found sufficient to provide the draught and carry away the smoke.

Construction.—One man excavates each trench, commencing from the ends nearest the chimney, another man cuts out the bottom of the chimney, and builds it up with the sods cut in construction of the trenches. The third man excavates the draught or flue, which is 12 feet long, 9 inches wide, and 6 inches deep; and as soon as the trenches are dug, he cuts a flue from the head of each into the main flue, taking care that the openings of the outer tunnels do not face one another (which would interfere with the proper working of the draught), then covers the flue with turf or sods from the top of the trenches to the chimney.

The other two men excavate the transverse trench, and provide turf for the construction of the chimney.

The men on the completion of the trenches are employed respectively in providing and mixing clay, carrying water, and covering the trenches for the reception of the kettles.

Great care must be taken in the construction of the chimney; all holes and interstices must be plastered with clay.

The insides of the trenches may be plastered with clay if it be plentiful. If this is done the dimensions should be slightly increased. If the clay is scarce the trenches should be cut smooth. Each trench will accommodate about 11 oval or 12 small oval kettles, the holes of which should be modelled in clay, using the base of a kettle as a mould. The intervals across the trench should be covered by sods placed grass side downwards, or stones, hoop-iron, sticks plastered with clay, and all interstices closed with clay or sods. This kitchen will last a fortnight even if not plastered with clay. Time required to construct 8 hours, working party one non-commissioned officer and twelve men, tools required:—

The British Army Cook Book 1914

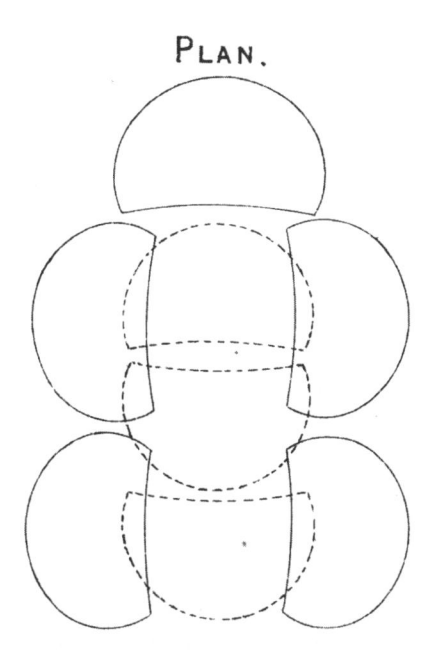

PLAN.

ELEVATION.

FIRE

Weller & Graham L^{td} Litho, London

Axes, pick	...	...	...	...	... 3
Hooks, bill	...	...	...	...	... 2
Kettles, camp	...	...	...	...	... 9
Pickets, bundle of		...	...	...	... 1
Spades	...	...	...	...	... 11

It will be seen that this kitchen admits of easy extension by the addition of more trenches.

Advantages.—More room is provided between the trenches for the cooks to work in, less time is required to build. 18 feet less ground is required to provide this transverse trench, the flues are easy to repair.

COOKING IN MESS TINS.

No trench should be dug; the mess-tins should be placed on the ground as shown on plan on opposite page, with the opening facing the direction of the wind.

Eight is a convenient number of tins to form a "kitchen," but any number from 3 to 10 or 11 can be utilised.

The handles of the mess-tins should be kept outside.

The tins should be well greased on the outside before being placed on the fire; if this is done and they are cleaned soon after being used they will suffer no damage. The tins when they are hot can be cleaned in a few minutes with turf, soil, or rag.

Only a small quantity of wood is required for each "kitchen," a good draught being the object to be kept in view. It is desirable that the fuel used should, whenever possible, be that obtainable in the vicinity of the "kitchens."

Each man should be instructed to cook his own dinner, but when once the "kitchen" is formed and the fuel collected one man only should remain with each fire.

The position of the tins in each "kitchen" will require to be changed from time to time, as some will be cooked sooner than others. It will be the duty of the man in charge to regulate this.

The dinners will be cooked from 1 hour to 1¼ hours.*

The following dinners are suitable for this method of cooking:—
Plain Stew, Irish Stew, Curried Stew, Sea Pies, and Meat Puddings.

For instruction it is convenient to place the "kitchens" of companies in rows at any convenient distance apart; but if space is limited, it is estimated that dinners of a battalion of 500 men can be arranged in a space of 10 yards by 16 yards, working by double companies, and allowing an interval of 2 feet between the "kitchens" of a company and 2 yards between the lines of each double company, or in a line of 32 yards by 4 yards, working by half battalions with the same intervals. When possible, the latter is the more suitable, as the men attending the fires are less inconvenienced by the smoke from the other "kitchens."

---

* Men's water bottles should be filled with water before quitting barracks.

PRESERVED MEAT TINS.

Preserved meat tins are a good substitute for either a kettle or canteen, and can be used for a variety of purposes. In opening them, care should be taken that the lid is not entirely removed, leaving about 1 inch uncut, it will then act as a lid. Another one should be cut open and emptied and then placed on the cinders a few minutes to melt the solder, then overlap it about 1 inch, which will decrease the circumference sufficiently for the first one to rest upon, then cut away on each side of the portion of the tin that was soldered a piece about 1 inch wide and 1½ inches deep, and overlap the two pieces in the form of a V; this will bind them together and form the flue. In the front at the lower part, a piece about 2 inches by 2 inches should be cut away to form an opening to feed the fire, which should consist of sprigs, rushes, turf, fir cones, or any small wood that may be procured. It will be found that by simply using these tins as described, it is possible to prepare tea, coffee puddings, pies, stews, rice, &c., in a manner equal to that produced by the kettle. The tins may also be used for preparing chuppatties, &c.

In a standing camp tubs are usually provided for the refuse, and the same conditions should attach to them as in barracks.

The rubbish pit is constructed by the sanitary squad of the unit, and is under the immediate care of the non-commissioned officer i/c of the squad.

The serjeant-cook should take a personal interest in the maintenance and care of the rubbish pit, and issue such orders to his men that only dry rubbish, food refuse and other dry material be placed in this pit in such manner as to keep the surroundings of the cooking place neat and tidy. Liquid refuse such as greasy water should not be placed in this rubbish pit, but thrown either into a special watertight receptacle, if available, or into a sullage pit, specially dug for the purpose by the sanitary squad men.

ALDERSHOT OVEN.

The Aldershot oven (Mark II) consists of:—

        2 sections.
        2 ends.
        1 bottom.
        4 bars.
        9 tins
        1 peel.
        Total weight 374 lbs. (about 3¼ cwt.).

The bottom can usually be dispensed with, in which case the above weight is reduced by 66 lbs.

The length of the 2 sections when up is 5 ft. 1 in.; width, 3 feet 6 inches.

Capacity.—Each oven will bake 54 2 lb. or 2½ lb. loaves (108 rations) in each batch, or if used for cooking, will cook dinners for about 220 men.

Time for Heating, Baking, Cooking, &c.

1st heating 1st day	4 hours.
1st heating 2nd day	2 hours.
2nd and subsequent heating	1½ hours.
Baking	1 to 1¼ hours.
Cooking	Up to 2¼ hours.

Fuel Required for each Oven.

1st heating 1st day	300 lbs wood.
1st heating 2nd day	150 lbs. wood.
2nd and subsequent heating	75 lbs (baking).
2nd and subsequent heating	Up to 150 lbs. (cooking).

A good rough rule for baking bread is to allow 1 lb. of wood for each pound of bread required.

Erecting the Oven.

Select a gentle slope on clay soil if possible, and avoiding marshy or sandy ground, the mouth of the oven to face the prevailing wind, as the oven heats more quickly and the smoke is blown away from the bakers or cooks.

The site should be cleared and smoothed, and sods should be cut to build up the back, front and sides of the oven. The bars are then placed over the site already prepared, the back one overlapping the front, the back of the oven placed in position, the plate forming the bottom of the oven is then placed against the front portion and firmly fixed, the sods are then built round the front, back and sides, a trench is next cut for the cook to work in, which is 18 inches deep, 2 feet wide and 6 feet long, leaving a space of 12 inches between it and the oven. The clay or soil from the trench being mixed with water and grass rushes, &c., to assist in binding it is then thrown on the oven and well beaten down. The depth of clay or earth should be at least six inches. The roof should slope backwards slightly, to carry off the rain.

Directions for Working.

1. Every night, wood should be laid in the oven ready for lighting in the morning. It is thus kept dry.
2. When the oven is heated, the embers are drawn out with a rake, and a small quantity of ashes left and raked even with the floor.
3. The tins containing the dough should not be put in till 20 minutes or half an hour after the fire is drawn, as otherwise the top heat is so fierce that it would burn the upper part of the bread.

When meat is to be cooked, it may be put in immediately the fire is drawn.

4. Immediately the oven is filled, the door should be put up and wedged tightly with a piece of wood, the end of which should rest on the outer edge of the trench in front.

The crevices round the end should then be filled in with wet clay to prevent any steam escaping. If this is properly performed the steam providing the necessary moisture is retained, and the bread or dinners will not be burnt.

5. When the top of the oven sinks to less than 14 inches from the bottom, which will happen sooner or later owing to the metal being softened by the heat, the oven should be taken to pieces and beaten into shape with mauls.

Ground ovens, on the principle of the Aldershot oven, may be improvised of almost any material, the most common being corrugated iron, barrels or half barrels, biscuit tins beaten out into the shape of an arch and supported by half tyres of wheels, &c. In each case the shape, method of working, &c., of the Aldershot oven should be taken as a pattern.

Ovens can be easily improvised, the main object being to obtain a covered-in space which will bear and retain the heat of a fire lighted inside. In a clay soil they can be dug out.

When using a field kitchen, a simple plan is to dig the transverse trench about 12 inches deeper in the part selected for the oven, then take the handle from the pick and drive it into the ground, measuring the distance from the edge of the trench, according to the length of the oven required, then dig out the soil or clay, leaving a portion for the roof about 8 inches deep, but this depends upon the ground. When the oven is sufficiently large, care being taken that it is not more than 12 inches high, the pick handle should be drawn out; the hole will then act as a flue. The fire should be lighted, and the oven allowed to become thoroughly hot before the dishes are placed in. A piece of turf large enough to cover the front of the oven should have been cut, which will act as the door, pugging it round as with the Aldershot oven.

Beer or biscuit barrels make excellent ovens, one end is knocked out, the ground slightly sloped, so that it may rest firmly, the sides, back, and top being covered with clay, well wedged downwards, to become quite hard; the fire is then lit and allowed to burn until the whole of the barrel is consumed; the hoops will then support the clay, and the oven may be safely used. Where the clay is good a small oven may be built by it alone. Build two walls the required distance apart, about 6 inches high, with clay that has been well beaten and mixed, the back being joined to the walls; then, with one hand on either side, gradually build the walls a few inches higher, the tops slightly sloping towards each other, leaving an interval in the form of a V in the centre, then mould a piece of the clay large enough to fill the space, and place it in, care being taken to well join the edges with the walls both inside and out; a small fire should then be lit and allowed to burn slowly until the clay is dry, it will then become baked and quite firm, and may be used as other ovens.

Tin biscuit boxes are also a good substitute for an oven. Melt one side of the solder, and form it into an oval shape, lay it on the ground, and cover with a few inches of clay or soil sufficient to retain the heat; light the fire, and proceed as with others.

Small joints of meat may be baked in the service camp kettle. When using a pugged trench a small amount of fat should be placed in the bottom, then a few clean pebbles large enough to cover the fat, the joint placed on the pebbles, and the lid put on. It requires a little longer to cook than the ordinary oven, and it is hardly possible to perceive any difference in the taste. Ant heaps are extensively used as ovens, the insides being scooped out and the fire lighted as in an ordinary oven.

KITCHEN.

The kitchen should at all times be kept as clean as possible.

On receipt of the wood, it should be built up in one or two stacks, according to the amount received, on either side of the chimney This will allow the front of the kitchen to be kept clear for the cooks to dish up, &c. After each meal the cook should understand that his first care is the trench. It must be kept clean, and nothing in the shape of wood, knives, &c., be allowed to remain upon it. In the evening the cook will, if necessary, repair his trench with clay or turf. He should then lay his fires as previously mentioned, and replace the kettles, the whole of the cooks assisting in the general clearing up. The intervals between the trenches should be swept downwards into the transverse trench, the front portion of the latter being swept upwards, it and the trench swept from end to end. No fuel should be wasted, wood only partly consumed, having been previously used, to lay the fires. The work must not be considered finished until everything is left in readiness for the morning.

WOOD.

The principal fuel used in the field is wood. It will at times be found necessary to use peat or turf, which has been before alluded to cow or horse dung; this should be mixed well with any rubbish grass, leaves, &c., and formed into convenient pieces for use in the trenches, and placed in the sun to dry. This is the principal fuel of the poorer classes in warm climates. On the River Nile the banks may be seen lined with this description of fuel drying in the sun. Fir cones and dried furze bushes are excellent for kindling purposes. In using wood it is necessary to cut it into short pieces, and split i lengthways, or otherwise it would become charred and retain its heat The daily allowance of wood at home is 3 lbs , on active service 2 lbs

Tommies heading for the Front.

Above: British Bovril – advertising with a hint of conscription.

Right: A poster showing a glimpse of life in the trenches.

WILL YOU MAKE A FOURTH ?

SAILORS' & SOLDIERS' TOBACCO FUN[D]

IT IS A SIGNIFICANT FACT THAT ALMOST EVERY LETTER FROM THE FRONT CONTAINS A REQUEST FOR "SOMETHING TO SMOKE"

Contributions gratefully received by
Hon. Sec., CENTRAL HOUSE, KINGSWAY, LONDON,

NATIONAL SERVICE
WOMEN'S LAND ARMY

"GOD SPEED THE PLOUGH
AND THE WOMAN WHO DRIVES IT"

PPLY FOR ENROLMENT FORMS AT YOUR NEAREST POST OFFICE OR EMPLOYMENT EXCHANGE

The Western Desert school of cookery.

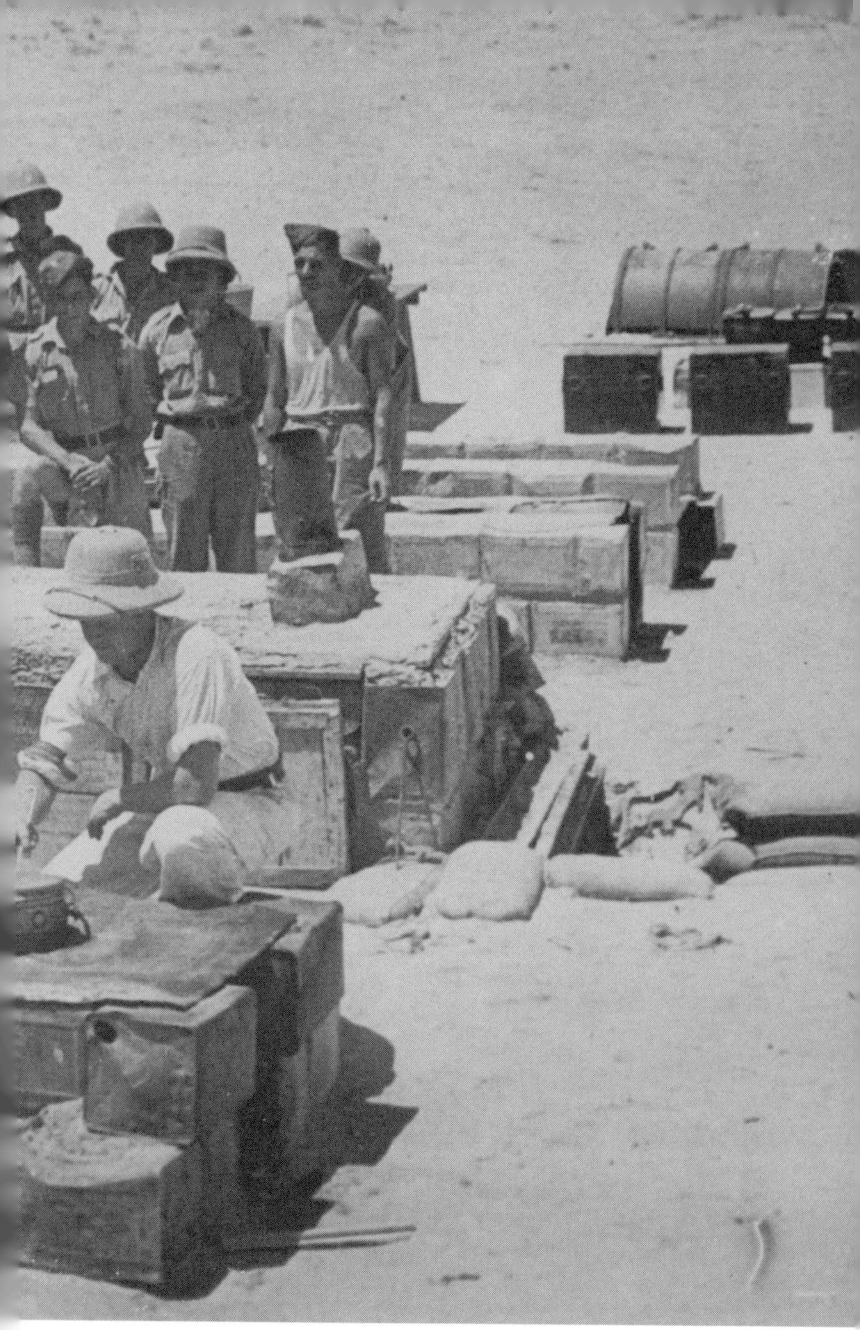

British machine gunners wearing gas helmets.

GAS HELMETS. OFFICIAL PHOTOGRAPH, CROWN COPYRIGHT RESERVED.

A glimpse of the devestation of the First World War.

La Grande Guerre 1914-15
de la rue Faidherbe après le bombardement   A. R.

368 GUERRE 1914-1915. — Conflit Européen.

Soldiers hunkering down in the trenches.

lgique. — *Avant-poste autour de Nieuport.*
Trenches before Nieuport. — LL.

O ALMOÇO NAS TRINCHEIRAS

Soldiers' daily rations were limited, but they made the most of them.

Serv. Phot. du C. E. Portugais. — Phot. Garcez.

LE DÉJEUNER DANS LES TRANCHÉES

Above: Soldiers taking rest in the trenches.

Below: Examining rations.

It wouldn't do for the general public or the enemy to know the true limits of rations on the Front Line, so propaganda often portrayed healthy, happy soldiers.

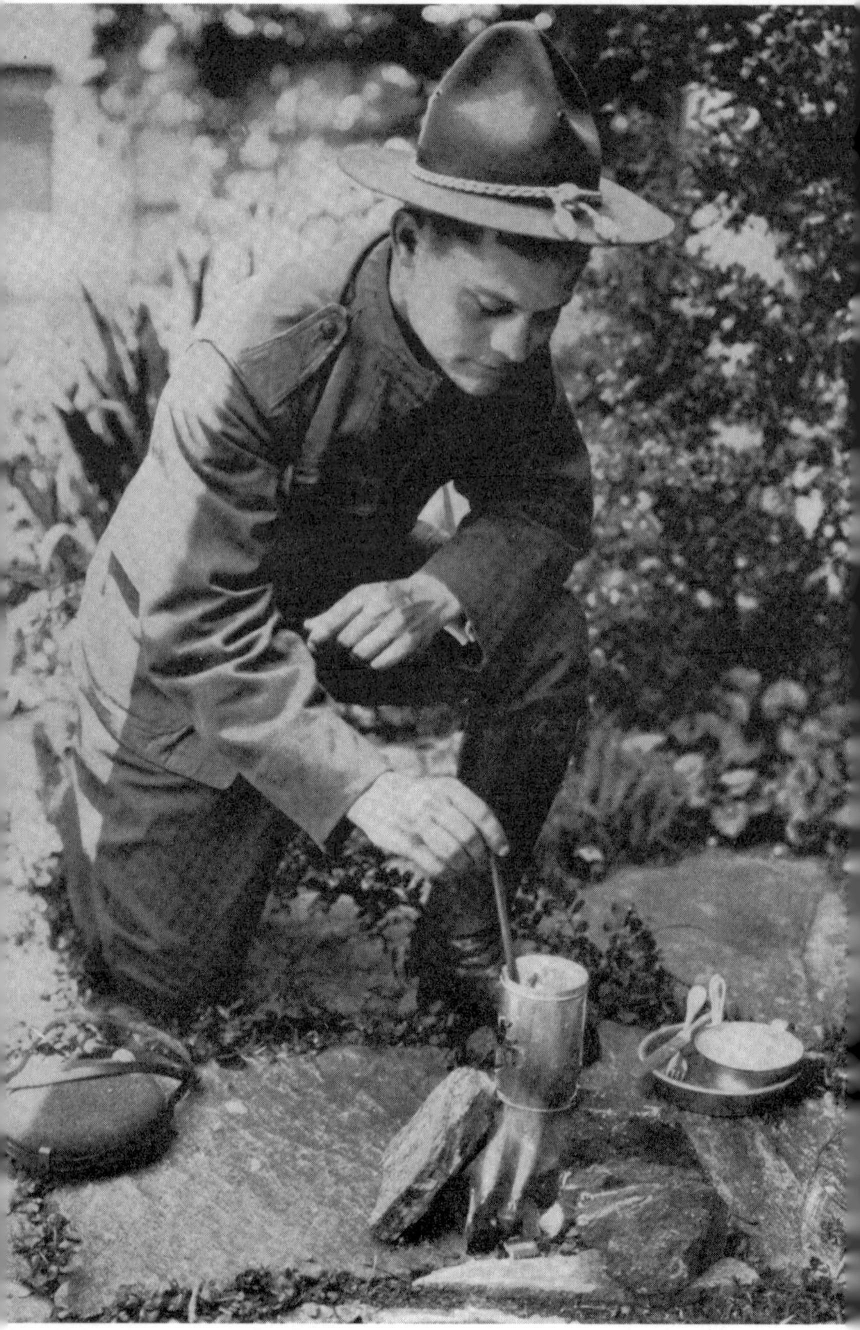

Opposite: A soldiers cooking up his rations.

Above: Soldiers learning French in the trenches on the Western Front.

Portioning out rations among the men.

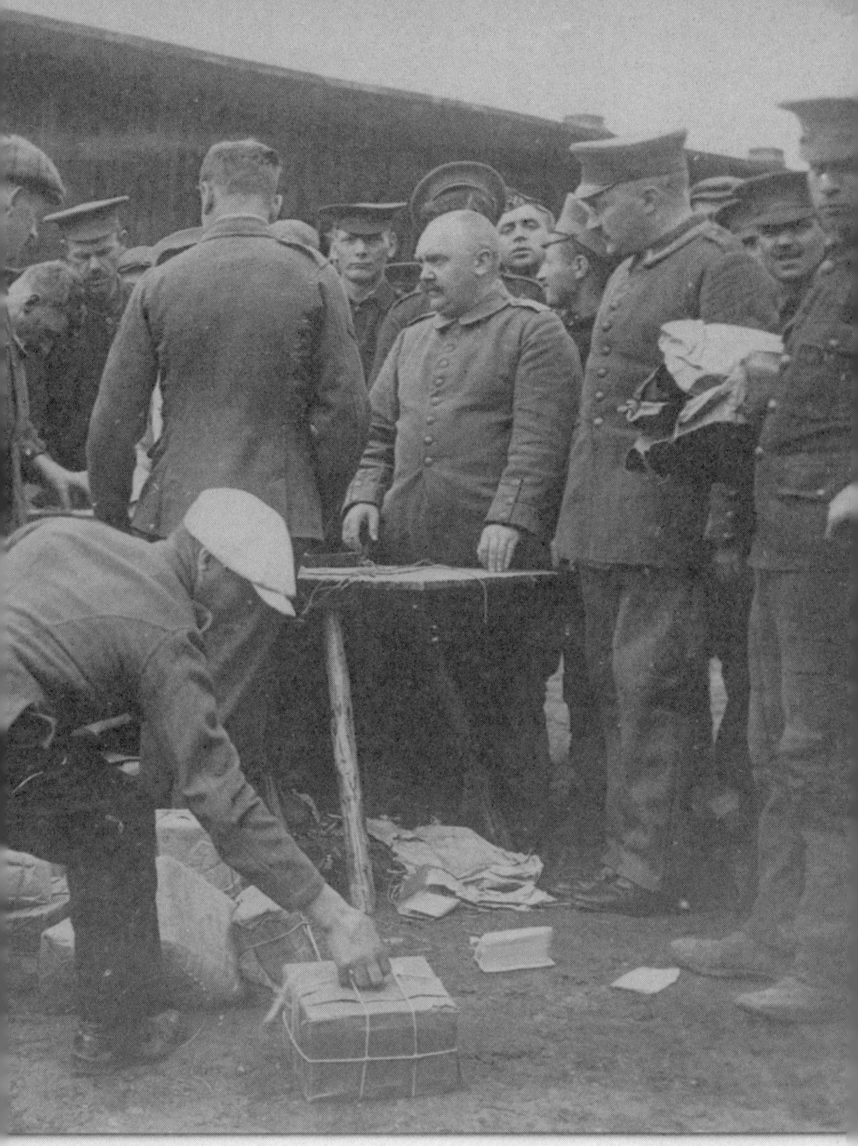

Troops arriving at Marlow.

Marlow July 2nd 1915

Staff at Hollingworth Lake, 1914.

AMP 1914. W. PERRY & Co. PHOTOGRAPHERS ROUNDACE

'Are we downhearted?'

"ARTED". ?NO"

An uncommon sight on the Front, perhaps not so further back behind the lines.

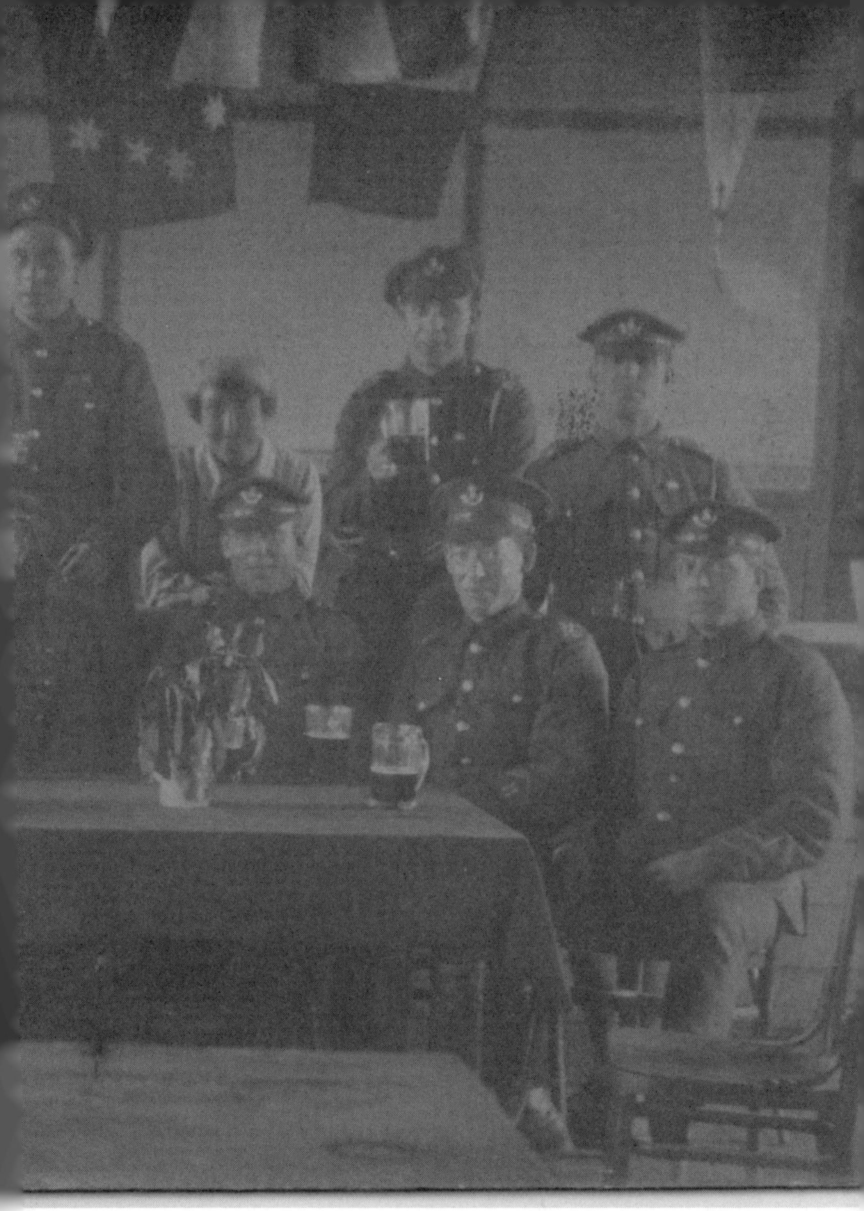

A German postcard of soldiers eating.

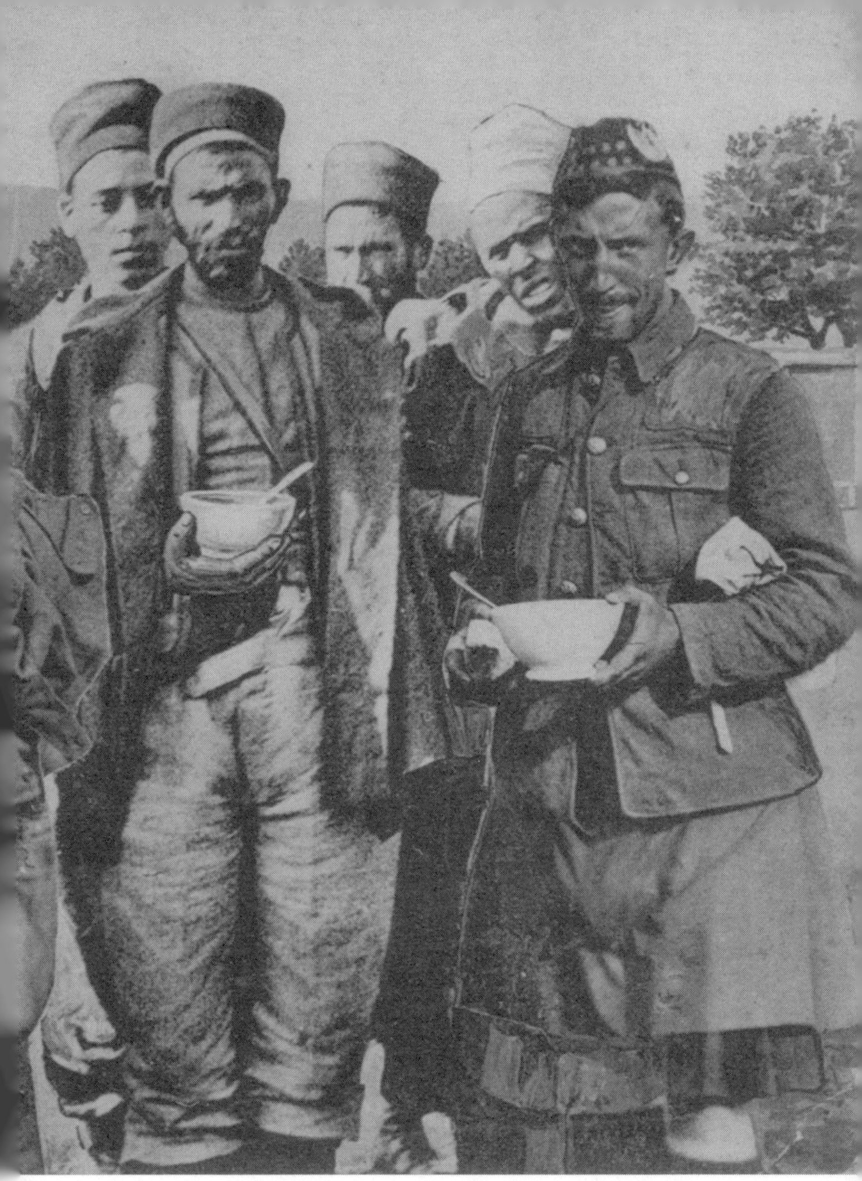

42. LA GUERRE de 1914 — Les Troup

Soldiers on the march.

nes en France — Indian troops in France

J. C, Paris

A view of the trenches.

MMY AT HOME IN GERMAN DUG-OUTS.

Happy Tommies wearing Hun helmets.

73. OFFICIAL PHOTOGRAPH,
CROWN COPYRIGHT RESERVED.

A slightly more gentile set-up.

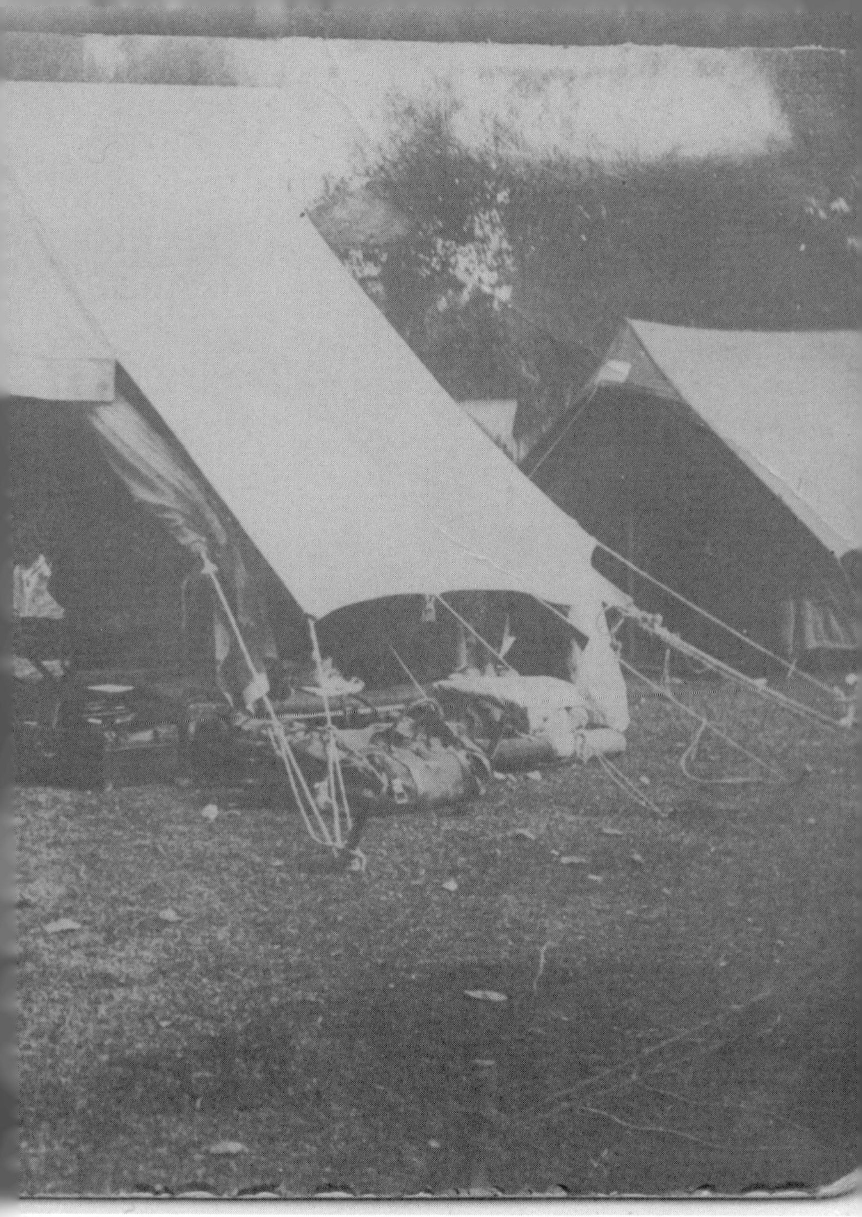

Serving up a meal.

17 - Guerre 1914 La Popote E.M.

Tommies in a captured German trench.

CAPTURED GERMAN VILLERS. Nº 12

RECIPES FOR FIELD COOKING.

PRESERVED MEAT.
(Ingredients for 22 Men.)

Meat Pie.

16¼ lbs. meat, 5 lbs. flour, 1½ lbs. suet, 1 lb. onions, 2 ozs. salt, ¼ oz. pepper.

Make the paste; cut up and stew the onions with jelly from the meat added; cut the meat into dice and place it in a baking dish; add the cooked onions; season with pepper and salt; cover with a light crust, and bake in a quick oven for 20 minutes.

Stew.

16¼ lbs. meat, 2 lbs. of carrots or other vegetables, 1 lb. onions, 2 ozs. salt, ½ oz. pepper.

Cut up the vegetables and onions, which place in the boiler with sufficient water to cover them; add some jelly from the meat; well season with pepper and salt; and stew gently, keeping the lid of the boiler closely shut until the vegetables are tender, then add the meat; let the whole simmer for 10 minutes and serve.

Curried Stew.

Ingredients the same as for stew, with 1 oz. of curry powder and 1 lb. of flour added. Prepare as for stew; mix the curry and flour with cold water into a smooth batter, and add it to the stewed vegetables with the meat; let the whole simmer for 10 minutes and serve.

Sea Pie.

Ingredients the same as for stew, with 5 lbs. of flour and 1½ lbs. of suet or dripping added.

Make the paste; prepare and cook the vegetables and onions, as for stew; when the vegetables are tender add the meat; cover the whole over with a light paste, and boil or steam for 20 minutes. A thickening of flour added is an improvement.

Toad-in-the-Hole.

16¼ lbs. meat, 5 lbs. flour, 1 lb. suet or dripping, 2 ozs. salt, ¼ oz. of pepper, 1 lb. onions, 6 eggs or equivalent in egg powder or 1 pint ale. Cut up and cook the onions; prepare the batter with eggs and milk if possible, if not with beer and water; season it with half the pepper and salt; grease the inside of a baking dish; pour into it half the batter, and place it in the oven; when the batter sets, place on the meat, cut up, and the cooked onions; cover with the remainder of the batter, and bake from 15 to 20 minutes in a quick oven.

POTATO PIE.

16¼ lbs. meat, 20 lbs. potatoes, 1 lb. onions, 3 ozs. salt, ½ oz. of pepper.

Cut up and stew the onions with jelly from the meat added; boil or steam the potatoes; when cooked mash them. Line the sides of the dish with one-third of the mashed potatoes; place the meat and cooked onions in the centre; season with pepper and salt; cover over the remainder of the mashed potatoes, and bake till the potato cover is brown. As the mashed potatoes absorb the moisture of the meat and render it dry, about 2 pints of gravy prepared from the liquor in which the onions were cooked, should be poured into the pie before serving.

HUNTER'S PIE.

This pie is prepared in a similar manner and with the same ingredients as potato pie, but the top is left uncovered. Both these pies should be baked in a quick oven.

The quantities of the several ingredients in the following recipes are not given, but they should be in the same proportion as the foregoing, which are for messes of 22 men each.

PEA SOUP WITH SALT PORK OR BEEF.

Meat, mixed vegetables, split peas, flour or broken biscuits, pepper, water.

Peel, clean, and cut up the vegetables; place the water in the camp kettle, add the vegetables and peas, and boil gently until the peas are soft. Then put into the soup about two lbs. of meat, which should have been previously well washed in cold water, and simmer gently till it is cooked, then take it out and cover it up to keep warm.

Mix some flour into a smooth batter with cold water, and add it to the soup, keeping it well stirred to prevent it burning; boil for 30 minutes, and serve. If flour is not to be had, use, instead, powdered biscuits, previously soaked in cold water.

The remainder of the meat should be soaked and well washed in cold water, then put in the camp kettle with sufficient water to cover it, and allowed to boil for 30 minutes; the water in which it was boiled should now be thrown away, the camp kettle refilled with fresh cold water, and the meat boiled till done.

IRISH STEW, WITH SALT BEEF.

Meat, potatoes, onions, and pepper. Wash and clean the meat in cold water, separate it from the bone, and cut it into small pieces of about 2 ozs. each, and well wash it again in cold water: peel and clean the potatoes, peel and slice the onions, place the meat, potatoes, and onions in the camp kettle, add a little pepper and sufficient cold water to cover the whole; put the lid on the kettle and cook gently over a slow fire, frequently skimming the fat off the top. The bones of the meat should not be added to the stew, as they are usually too salt.

Salt Pork and Biscuit.

Meat, biscuit, onions, parsley, pepper, and water.

(*a*) Soak the biscuits in cold water for 1 hour, wash clean, and boil the pork; drain the water off the biscuits, and cut up the pork into thin slices; peel and slice the onions, wash and chop up the parsley, pour a little water into the camp kettle, place a layer of the slices of pork at the bottom of the kettle, with some onions, parsley, and pepper, then a layer of the soaked biscuits on top, then a layer of pork, and so on alternately until the kettle is nearly full. Cover the whole with water, and cook gently over a slow fire for 1 hour and 15 minutes, and serve.

(*b*) Treat the pork, onions, and parsley as in (*a*). Soak the biscuits for 2 hours, then squeeze them dry, mince up the pork and mix it with the biscuits, onions, parsley, and pepper; then roll it into balls, and place in a camp kettle, with sufficient water to cover, and cook gently over a slow fire, and serve.

These recipes can also be prepared in the camp kettle lids by placing the layers of pork and biscuits or balls in one camp kettle lid, and covering it with another, and placing a few live embers underneath and on top of the lids.

Salt Beef and Dumplings.

Meat, flour, suet, water. Soak and well wash the meat in cold water, and place it in the camp kettle with plenty of water, and boil gently for 1 hour; then throw away the water in which it was boiled, and replace it with fresh cold water, and boil till the meat is cooked. Chop the suet up fine, mix it with the flour, and pour in some cold water and well mix the whole, and form it into dumplings about 2 inches in diameter; place the dumplings in the kettle with the beef about 30 minutes before the latter is cooked, and let both boil together until done.

Soup with Australian Preserved Meat.

Meat, mixed vegetables, flour, pepper, salt, barley, water.

Place the water in the camp kettle, scrape and clean the vegetables, add them to the cold water; when the water boils, shake in the dry barley. When the vegetables are cooked, take them out and pulp them; mix the flour into a smooth batter with cold water, add it to the vegetables with salt and pepper, and put the whole into the camp kettle, keeping it well stirred to prevent burning; allow it to simmer gently for 30 minutes, then open the tins of meat and add the contents to the soup, stir well, and simmer for ten minutes, and serve.

Stew with Australian Preserved Meat.

Meat, potatoes, onions, pepper, salt, and water.

After preparing the onions and potatoes put them in the camp kettle, season with pepper and salt, pour in sufficient water to cover them, and stew gently, keeping the lid of the vessel closely shut until the potatoes are nearly cooked; then open the tins of meat and cut up the contents, and put them in the kettle with the potatoes, and let the whole simmer for 10 minutes, then serve.

BROWN STEW WITH AUSTRALIAN PRESERVED MEAT.

Peel and slice the onions, melt the fat of the meat in the camp kettle, add the onions and fry them till brown, mix the flour into a smooth batter with cold water, season with pepper and salt, and pour it into the camp kettle, stir the whole well together, cut up the meat into slices, put it into the kettle, and when warmed through serve.

Directions as to the best methods of treating the rations when the rations are to be carried by the men, will be found on Schedule VI.

POINTS TO BE NOTED ON VISITING A KITCHEN.

1. Ask for weekly diet sheets and compare the dinners with the description shown.
2. Compare grocery books with the diet sheets; the description of each meal should agree.
3. See the articles necessary to provide the dish have been paid for by the company.
4. Open the whole of the grocery books for any given day and compare the diets shown with the articles purchased; by doing this you can form a good opinion of how the messing of the various companies has been carried out.
5. Note the prices charged for various articles, especially those required frequently, such as tea, flour, sugar, onions, potatoes, mixed vegetables, milk, &c.
6. Note the stock pot: bones broken small to extract their value; pot or boiler to be frequently skimmed and the stock to be used in lieu of water in preparing dinners when the whole of it is not required for soup Gravy should always be prepared from the stock pot.
7. See the dripping itself and dripping return; note the disposal of the dripping and how accounted for.
8. Inspect the grocery cupboards and see if the contents agree with the articles that have been paid for.
9. Note the general appearance of the food, cooks, and kitchen; if you do not find cleanliness you cannot expect very favourable results.
10. Note recapitulation in each book
11. Excepting Numbers 5 and 10, the serjeant-cook should be held responsible for any mistake that has not been brought to the notice of the Officer commanding the company.

APPENDIX.

SCHEDULES.

Illustrative of the Present System of Military Cooking, showing Economy Effected, the Variety of Diet Obtained and the Alimentary Principles Necessary for a Soldier or Working Man.

The following Schedules are published in order to illustrate the principles laid down in the " Memoranda upon the Messing of the Soldier," issued in May, 1891, and revised, August 1st, 1892. They form a sequel to the above memoranda, and show the economy effected, the variety of diet obtained, and the alimentary principles necessary for a soldier or working man, under the several conditions of every day life.

The system advocated by the Army School of Cookery is one which is mainly applicable to the feeding of, and cooking for, large numbers of individuals, as for example, in public institutions, schools, colleges, ships, in H.M. Army and Navy. The prices of groceries, vegetables, meat, and bread are those which can only be obtained by contract for supply on a considerable scale, and in judging of the prices given in the following Schedules, this fact must be taken into consideration.

Schedule I. Variety of diets and ingredients required.

II. Specimen of Military Diet for a unit in camp.

III. Specimen of Military Diet for one week of a battery, squadron or company in barracks.

IV. Monthly return of Dripping.

V. Details of saving of Stock and Dripping.

VI. Methods of treating rations when to be carried by the man.

SCHEDULE 1.

The following is given to assist in making up and checking Army Book B 48:—

DINNERS.

A
1.—Baked meat and potatoes.
2.—Baked meat and haricot beans.
3.—Baked meat and lentils or peas.
4.—Roast meat and Yorkshire pudding.
5.—Roast meat stuffed.
6.—Brown curry and rice.
7.—Meat pies.
8.—Brown stew.
9.—Plain „
10.—Irish „
11.—Curried „
12.—Meat steamed.
13.—Meat steamed with peas.
14.—Meat steamed with haricot beans.
15.—Meat puddings.
16.—Sea pies.
17.—Turkish Pillau.
18.—Toad-in-the-hole.

B
1.—Barley soup.
2.—Pea „
3.—Lentil „
4.—Lentil and pea soup.
5.—Hotch potch.

C
1.—Plain pudding.
2.—Plum „
3.—Jam rolls.
4.—Currant rolls.
5.—Plain raisin pudding.
6.—Date pudding.
7.—Rice pudding.
8.—Bread and butter pudding.
9.—Tapioca.
10.—Apple, rhubarb, or other fruit tarts.
11.—Apple, rhubarb, or other fruit puddings.
12.—Treacle pudding.
13.—Macaroni.

It is optional which dinners are selected, but soups should always be given with a roast or bake, the ovens and boilers being used alternately by the various companies.

BREAKFAST OR TEA.

Butter.
Porridge and milk.
    „ with golden syrup, jam or sugar.
Cheese or Welsh rarebit.
Fried liver.
    „   „ and bacon.
Boiled eggs.
Fried eggs and bacon.
Bacon fried or boiled.
Brawn.
Faggots or rissoles.
Curried liver.
    „   brawn.
Tripe, stewed or curried.

Corned meat.
Bloaters.
Haddocks.
Kippers.
Sprats, pickled or fried.
Herrings, pickled or fried.
Fish cakes.

WHEN IN SEASON.

Lettuce.
Spring onions.
Radishes.
Cresses.
Plain salad.
Dressed salad.

SCALE OF INGREDIENTS REQUIRED FOR 60 MEN.

1.—Baked meat and potatoes—potatoes 5 stone, onions 2 lbs., pepper ½ oz., salt 1 oz.
2.—Baked meat and haricot beans—onions 2 lbs., haricot beans 8 lbs., pepper ½ oz., salt 1½ ozs.
3 —Baked Meat and blue peas—onions 2 lbs., blue peas 8 lbs., pepper ½ oz., salt 1½ ozs.
4.—Roast meat and Yorkshire pudding—flour 10 lbs., egg powder 6 packets, milk 5 quarts, pepper ¼ oz., salt 1 oz.
5—Roast meat stuffed—onions 2 lbs., bread 6 lbs., parsley small bunch, eggs 4, pepper ½ oz., salt 1 oz.
6.—Brown curry and rice—mixed vegetables 4 lbs., onions 2 lbs., flour 2 lbs., curry powder 8 oz., pepper ½ oz., salt 1½ oz., rice 6 lbs.
7.—Meat pies—onions 2 lbs., flour 12 lb., pepper ½ oz., salt 1 oz.
8.—Brown stews—mixed vegetables 4 lbs., onions 2 lbs., flour 2 lbs., pepper ½ oz., salt 1 oz.
9.—Plain stew—mixed vegetables 4 lbs., onions 2 lbs., flour 2 lbs., pepper ½ oz., salt 1 oz.
10.—Irish stew—potatoes 5 stone, onions 4 lbs., pepper ½ oz., salt 1 oz.
11.—Curried stew—onions 2 lbs., mixed vegetables 4 lbs., flour 2 lbs., curry 8 ozs., pepper ½ oz., salt 1½ ozs.
12.—Steamed meat—mixed vegetables 4 lbs., onions 2 lbs., pepper ¼ oz., salt 1½ ozs.
13.—Steamed meat with peas—onions 2 lbs., blue peas 8 lbs., pepper ½ oz., salt 1½ ozs.
14.—Steamed meat with haricot beans—onions 2 lbs., haricot beans 8 lbs., pepper ½ oz., salt 1½ ozs.
15 —Meat puddings—flour 12 lbs., onions 2 lbs., pepper ½ oz., Salt 1½ ozs.
16.—Sea pies—flour 12 lbs., potatoes 5 stone, mixed vegetables 2 lbs., onions 2 lbs., pepper ½ oz., salt 1½ ozs.
17.—Turkish pillau—rice 6 lbs., cayenne pepper ¼ oz., onions 2 lbs. Salt 2 ozs., sweet herbs 1 bunch, flour 2 lbs.
18.—Toad-in-the-hole—ingredients as for No. 1, with 2 lbs. onions, 4 quarts milk.

1.—Barley soup—barley 6 lbs., mixed vegetables 6 lbs., onions 2 lbs., flour 2 lbs., celery seed 1 packet, pepper ½ oz., salt 2 ozs.
2.—Pea soup—split peas 7 lbs., mixed vegetables 6 lbs., onions 2 lbs., flour 2 lbs., dried mint 1 packet, pepper 1½ oz., salt 3 ozs.
3—Lentil soup—lentils 6 lbs., mixed vegetables 6 lbs., onions 2 lbs., flour 2 lbs., herbs 1 packet, pepper ½ oz., salt 3 ozs.

4.—Pea and lentil soup—lentils 4 lbs., split peas 3 lbs., mixed vegetables 6 lbs., onions 2 lbs., flour 2 lbs., pepper 1¼ ozs., salt 6 ozs., mixed herbs ½ packet.
5.—Hotch potch—blue peas 4 lbs., barley 3 lbs., mixed vegetables 6 lbs., onions 2 lbs., flour 2 lbs., pepper ½ oz., salt 3 ozs., cabbage lettuces or cabbages 6 heads, packet of sweet herbs, and small bunch of parsley.

1.—Plain suet pudding—flour 15 lbs., salt 1 oz., baking powder 2 packets.
2.—Plum pudding—flour 15 lbs., raisins 3 lbs., currants 3 lbs., salt 3 ozs., baking powder 2 packets, treacle 1 lb., or sugar 1 lb., spice 1 packet.
3.—Jam rolls—flour 15 lbs., jam 6 lbs., baking powder 2 packets, salt 1 oz.
4.—Currant rolls—flour 15 lbs., currants 6 lbs., baking powder 2 packets, salt 2 ozs., sugar 2 lbs., 1 lb. of lemon peel.
5.—Plain raisin pudding—flour 15 lbs., raisins 6 lbs., sugar 2 lbs., baking powder 2 packets, salt 1 oz.
6.—Date pudding—dates 12 lbs., flour 10 lbs., sugar 3 lbs., salt 1 oz., 1 nutmeg.
7.—Rice pudding—rice 6 lbs., milk 3 gallons, nutmegs 3, sugar 3 lbs., butter ½ lb.
8.—Bread and butter pudding—sliced bread 12 lbs., sugar 3 lbs., currants 3 lbs., butter 1½ lbs., milk 3 gallons, 1 nutmeg.
9.—Tapioca pudding—tapioca 6 lbs., milk 3 gallons, sugar 3 lbs., nutmegs 3, butter ½ lb.
10.—Apple tarts—flour 10 lbs., apples 15 lbs., sugar 3 lbs., cloves ½ oz., salt 1 oz.
11.—Apple puddings—as above.
12.—Treacle pudding—flour 15 lbs., treacle 6 lbs., salt 1 oz.

76

SCHEDU

FIELD CO

SPECIMEN OF MILITARY DIET IN A

Being the Diet of three Companies of Mounted Infantry encamped

GOVERNMENT RATIONS.

Meat, 16 ozs., including bone
Bread, 16 ozs.
} Net value, 6½d.

MESSING

3d. per diem deducted fro Soldier.

Total nett cost per man, per diem, inclu

NOTES.

The Dripping saved during the month of July, 1891, was 9,536 ozs., or 596 lbs., valued at £9 18s. 0d., being at the rate of £115 per annum for 374 men.

One lb. of meat and bone produced nearly 1 oz. of Dripping in addition to Stock.

A tent or shelter for cooks must be provided.

MILITARY F

Under the new sy

Nos. V., IX., and X. COMPAN

REGIMENTAL FIEI

SCALE OF DIET FOR WEEK E

Troop, Battery, or Company.	Approximate No. in Mess.	MEALS.	SUNDAY.	MONDAY.	TUESDAY.
No. V.	130	Bkfst.—½Co. ,, ½Co. Dinner Tea	Coffee, Bacon, and Steaks. Baked Meat and Potatoes. Plum Pudding. Tea and Dripping.	Coffee, Steaks, and Bacon. Brown Curry, Rice and Potatoes. Tea and Marmalade.	Coffee, Liver a Bacon, and Stea Roast Meat, Po toes, and Yorksh Pudding. Tea and Dripping.
No. IX.	124	Bkfst.—½Co. ,, ½Co. Dinner Tea	Coffee, Bacon, and Stews. Baked Meat and Potatoes. Plum Pudding. Tea and Dripping.	Coffee, Roast Meat, and Salmon. Meat Puddings and Potatoes. Tea and Jam.	Coffee, Stew, a Liver. Baked Meat a Potatoes. Pl Suet Pudding. Tea and Dripping
No. X.	120	Bkfst.—½Co. ,, ½Co. Dinner Tea	Coffee, Bacon, and Beef. Baked Meat and Potatoes. Jam Rolls. Tea and Dripping.	Coffee, Beef, and Brawn. Irish Stew. Tea and Marmalade.	Coffee, Brawn, a Beef. Roast Meat, Yo shire Pudding Potatoes. Tea and Dripping

Bourley Camp, Aldershot.—Date, 20th August, 1891.
(3563)

E II.

...KING.

...ANDING CAMP FOR ONE WEEK.
Bourley, near Aldershot, for the week ending 5th Sept., 1891.

...NEY.		APPARATUS AND FUEL.
...aily pay of each	Broad Arrow Kitchen. Aldershot Oven. Soyer's Stoves for Stock.	Service Kettle, 12 quarts. ,, ,, 7 ,, 3 lbs. of wood per man per diem.
...; fuel	9¾d.	

...D DIET.

...n of cooking.

..., MOUNTED INFANTRY.

...DIET RETURN.

...NG 5TH SEPTEMBER, 1891.

REMARKS.

The extra ¼ lb. of meat allowed for troops under canvas is used to provide a meat breakfast. Alternate companies or half-companies.

The staff of cooks required are—1 Serjeant-Cook, 1 assistant to keep Stock and Dripping (with returns), and 1 cook and 1 assistant per company.

WEDNESDAY.	THURSDAY.	FRIDAY.	SATURDAY.
Coffee, Steaks, Liver and Bacon.	Coffee, Liver and Bacon, and Steaks.	Coffee, Eggs and Bacon. Liver—Bacon and Steaks.	Coffee, Liver. Bacon, Steaks—and Eggs and Bacon.
Meat Pies and Potatoes.	Baked Meat, Potatoes and Pea Soup.	Curried Stew and Potatoes.	Roast Meat (Stuffed), Potatoes and Barley Soup.
Tea and Dripping.	Tea and Jam.	Tea and Dripping.	Tea and Fried Fish.
Coffee, Bacon, and Roast Meat.	Coffee, Curried Liver, and Stew.	Coffee, Roast Meat, and Brawn.	Coffee, Stew, and Bacon.
Baked Meat, Haricot Beans and Potatoes.	Roast Meat, Yorkshire Pudding and Potatoes.	Baked Meat and Potatoes. Plum Pudding.	Sea Pies.
Tea and Fresh Fish.	Tea and Dripping.	Tea and Kippers.	Tea and Syrup.
Coffee, Beef, and Butter.	Coffee, Butter and Beef.	Coffee, Beef, and Fried Liver.	Coffee, Fried Liver, and Beef.
Brown Curry, Rice and Potatoes.	Baked Meat, Haricot Beans and Potatoes.	Brown Stew and Potatoes. Currant Rolls.	Baked Meat, Potatoes and Pea Soup.
Tea and Fried Fresh Fish.	Tea and Cheese.	Tea and Dripping.	Tea and Dripping.

RAYMOND PORTAL, Capt & Adjt., Mounted Infantry

SCHE[

SPECIMEN OF A BATTERY, SQU[

MEALS.	SUNDAY.	MONDAY.	TUESDAY.	WEDN[
Breakfast...	Tea, Coffee, or Cocoa, Fried Bacon.	Tea, Coffee, or Cocoa, Butter.	Tea, Coffee, or Cocoa, Haddocks.	Tea, Co Cocoa,
Dinner ...	Baked Meat, with Potatoes and Apple Tarts.	Irish Stew, and Rice Pudding.	Roast Meat, Stuffed, and Pea Soup.	Meat and Ca
Tea ...	Tea, Bread and Butter.	Tea and Marmalade	Tea, Butter, and Salad.	Tea, I and C

Things to note when co[

1. The diet must be good and varied, no dish to be serv[
2. No two dishes containing the same ingredients shoul[Beans or Peas, Meat Pies and a pudding made of flour, Ric[
3. Green and other vegetables should be given in fair p[
4. The diet should be so arranged that the men may h[
5. Following points must be considered:—Price of mat[times when fish, &c., are in season, as these are sometimes

ULE III.

...RON, OR COMPANY DIET SHEET.

...AY.	THURSDAY.	FRIDAY.	SATURDAY.	REMARKS
..., or ...ter.	Tea, Coffee, or Cocoa, Sausages and Mashed Potatoes.	Tea, Coffee, or Cocoa, Butter.	Tea, Coffee, or Cocoa, Boiled Bacon (Cold).	
...es ...ge.	Roast Meat. Potatoes and Plum Pudding.	Turkish Pillau and Cabbage.	Tomato Soup, Roast Meat, Yorkshire Pudding, Potatoes.	
...er, ...se.	Tea, Butter.	Tea and Dripping.	Tea and Butter.	

...*ling a weekly Diet Sheet.*

...twice during the week.

...e served at one meal, *e.g.*, Pea or Lentil Soup, and Haricot
...th Curry and Tapioca, Rice or Sago Pudding.

...ortion; also Salads when in season.

...a roasting joint one day and a stewing joint next, and so on.

...ls, the money to be expended, the tastes of the men, and the
...aper when plentiful.

SCHE

MONTHLY RE

Return illustrative of Saving in

1st BATT. NORTHA

QUARTERMASTER'S DRIPPING RETURN

MONTHLY BALAN

Date.	Average No. in Mess.	Dripping saved during Month.		Issued Free to Companies.	
		lbs.	ozs.	lbs.	ozs.
1st to 31st March, 1892	843	781	...	643	...

1st *April*, 1892.
    Aldershot.

        Total value of Dripping saved a
        Value per annum at   ...   ...

JLE IV.

RN OF DRIPPING.

pping by Present System of Cookery.

TONSHIRE REGIMENT.

(*Appendix D.*)

SHEET, MARCH, 1892.

urplus Sold.		Value at 4*d.* per lb.			Remaining on Hand.		Remarks.
bs.	ozs.	£	s.	*d.*	lbs.	ozs.	
100	...	1	13	4	38	...	

J. *DEVLIN, Lieut. and Quartermaster,*
                        1st *Northampton Regiment.*

ove £13 0 4
.. .. £156 4 0

SCHEDULE V.

SAVING OF STOCK AND DRIPPING.

The following details show the Comparative Economy and saving of Dripping and Stock effected by the present system of Cookery.

100 lbs. ration meat or 133⅓ rations for 133 men, are found to produce as follows :—

Stock.

Stock	6½ gallons
Fat skimmed and clarified	1 9/16 lbs.
Average amount of bone	18¼ lbs.

Dripping.

Suet or surplus fat removed previous to cooking and clarified	3 2/16 lbs.
Fat skimmed during cooking and clarified	11/16 lbs.
Fat skimmed from stock-pot and clarified	1 9/16 lbs.

	s.	d.
Sale of bones at 2s. per cwt. realised ..	0	3¼
Value of dripping at 4d. per lb.	1	9½
Total	2	0¾

It is found preferable to retain the dripping made from the suet for issue in lieu of butter, and that made from skimmings of the stock-pot for cooking purposes.

The results above described are the results of experiments conducted at the Army School of Cookery with the ration meat issued to the troops at Aldershot.

SCHEDULE VI.

The following is considered the best method of treating rations when the rations are to be carried on the man.

The chief essentials in view are:—

1. That rations should be prepared so as to prevent them going bad before they are required for consumption.
2. That the rations should be so divided that their carriage by the soldier entails an even distribution of weight on each individual.
3. That the preparation of the rations before issue should be advanced to such a state that their subsequent treatment before they are ready for consumption is reduced to a minimum.
4. That the rations, as carried by the soldier, should be capable either of being cooked in the mess-tin or collected and dealt with by the company cooks.

The following methods of dealing with rations are suggested:—

(A). Method of dealing with full meat ration.

Breakfast (stew); Mid-day meal (haversack ration of bread and cheese); Evening meal (steak).

On receipt of the meat ration, it should be cut up into portions corresponding to the number of squadrons, troops, or companies.

The company cooks should then remove the meat from the bone.

The best portion of the meat should then be selected and cut up into thin steaks consisting as near as possible of portions each weighing 6 ozs. If the weather is hot these steaks should be fried in fat until partially cooked. They should then be allowed to cool. The remainder of the meat should be prepared for making stew. The stew should be cooked in the morning and issued for breakfast. The steaks should be laid on two clean waterproof sheets, and the company should be formed up in two ranks with their mess-tins. Both ranks should then file past, each man receiving his ration of meat and the necessary condiments.

For the mid-day meal the haversack ration of bread and cheese should be utilised.

On arrival in camp, if it is proposed that the men should cook their own dinner, all that is necessary is to add a little water to their mess-tins, stir well and go on with the cooking; or the meat can be fried if sufficient fat is available. Otherwise the meat can be dealt with as follows:

One camp kettle per fifteen men, containing a little water, should be placed on the flanks of the company; the men should then file past emptying the contents of their mess-tins into the kettles. The company cooks can then deal with the meat in the ordinary way.

This method would save time, besides leaving the men free for other work.

(B). Method of dealing with fresh meat rations.

Breakfast (fried steaks); Mid-day meal (haversack ration of bread and cheese); Evening meal (stew).

As in (A), the company cooks should cut up the meat. The best portions should be cut into steaks, each weighing as near as possible 6 ozs. These steaks should be fried and issued for breakfast. The remainder of the meat should be cut into small cubes, sprinkled with flour, pepper, and salt, and placed on clean waterproof sheets and divided up into portions. The company should then file past with their mess tins, each man receiving his portion together with a piece of onion. The meat can be collected on arrival in camp and dealt with by the company cooks, or, if desirable, the men can cook the meat in their mess tins.

In the latter case, the men should add enough water to barely cover the meat, stir well, and cook over a small fire for about 1½ hours.

(C). Method of dealing with meat rations when they consist of half fresh meat and half preserved meat.

When the meat ration consists of half fresh meat and half preserved meat, the preserved meat can be divided into two portions. (1) for breakfast, (2) for the mid-day meal; the half ration of fresh meat being prepared by company cooks, either for frying (when suitable portions are available), or for stewing. On arrival in camp, the fresh meat can either be cooked by the men in their mess tins or collected and dealt with by the company cooks, *or* the fresh meat can be made into stew or cut into steaks and fried for breakfast, the preserved meat being issued to the men, one portion being used for the mid-day meal and the remainder for the evening meal. The latter could be heated up if desired. By adopting the former method men are enabled to obtain their evening meal without delay, and without the necessity of waiting while the meat is cooking; on the other hand, it is desirable to give men a hot evening meal when possible.

When both fresh meat and preserved meat is carried in the mess tin, the former should be placed at the bottom and well pressed down. The preserved meat should be put on top.

When preserved meat is issued in small tins, it will be unnecessary to remove the meat from the tins until required.

Preparation of haversack rations, consisting of meat, for use when bread and cheese is not available:—

Take one-third of the issue of meat, cut into joints, and boil them until tender, which will take about 2½ hours. Remove from the liquor and allow to cool. Cut into thin slices and issue. The remaining two-thirds of the issue can be treated as already described for stews, etc.

STEAKS.

Take the best portions of the meat and cut into slices about 4 ozs. each, fry them in fat, and while warm press between two slices of bread, and issue as a haversack ration.

Hints for cooking in Mess Tins.

A mess tin will cook one man's meat with vegetables, and two men's without vegetables.

Small fires only are required, as rapid boiling makes meat tough and hard.

If possible grease the tins on the outside before placing them on the fire ; this makes them easier to clean afterwards.

Not more than 10 mess tins should be placed around one fire.

Utensils available for Dealing with Rations.

The utensils available for cooking in camp are limited, and as the carriage of individuals entails additional use of the cooking vessels, some arrangement is necessary.

The opportunity should be taken, when it is decided to partially cook the rations before issue, to do this over night, leaving the kettles free for the preparation of breakfast on the following morning. When the evening meat rations are to be dealt with by company cooks, it will often be found advisable to make the men prepare their own tea, and *vice versâ.*

ALSO AVAILABLE FROM AMBERLEY PUBLISHING

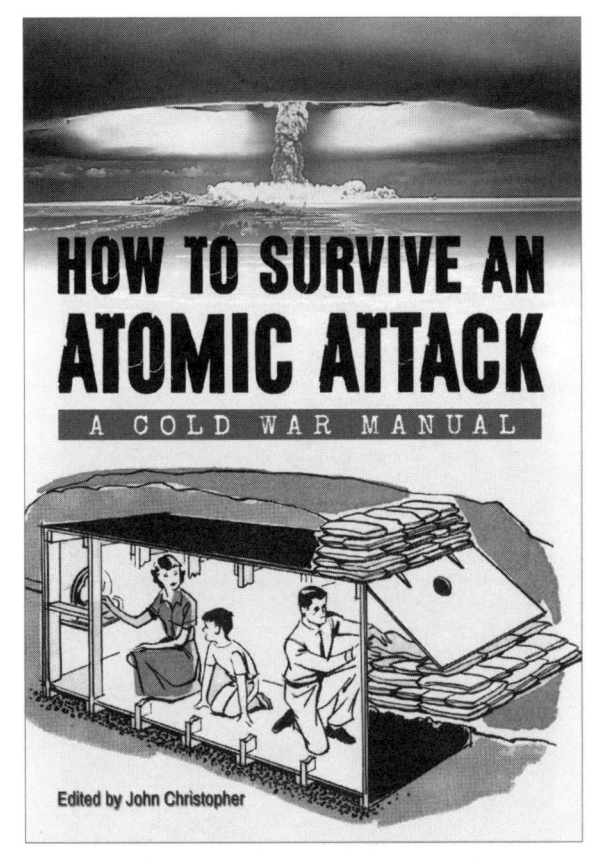

How to Survive and Atomic Attack
Edited by John Christopher

Official US government advice on how to survive an atomic attack, including full instructions on building your own nuclear shelter.

978 1 4456 3997 0
96 pages, full colour

Available from all good bookshops or order direct
from our website www.amberley-books.com

ALSO AVAILABLE FROM AMBERLEY PUBLISHING

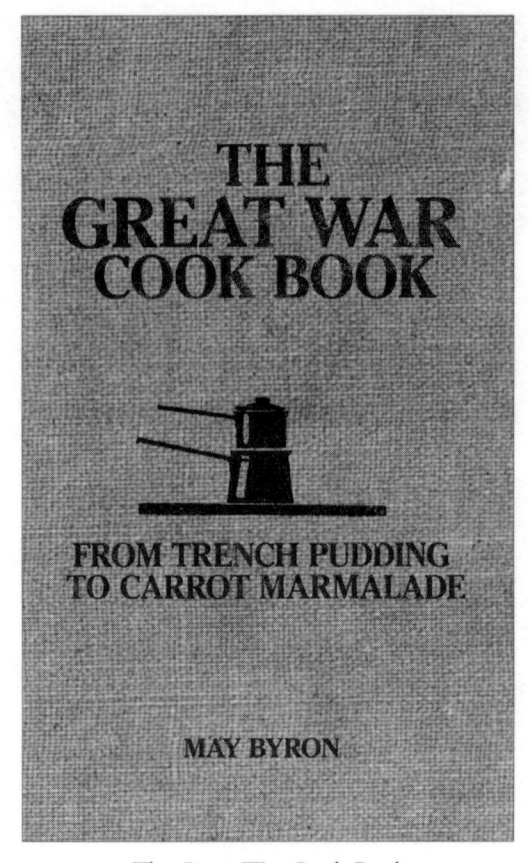

The Great War Cook Book
May Byron

With over 500 wartime recipes, May Byron offers unusual alternatives to traditional ingredients in a Britain almost starved into submission.

978 1 4456 3388 6
231 pages

Available from all good bookshops or order direct
from our website www.amberley-books.com

ALSO AVAILABLE FROM AMBERLEY PUBLISHING

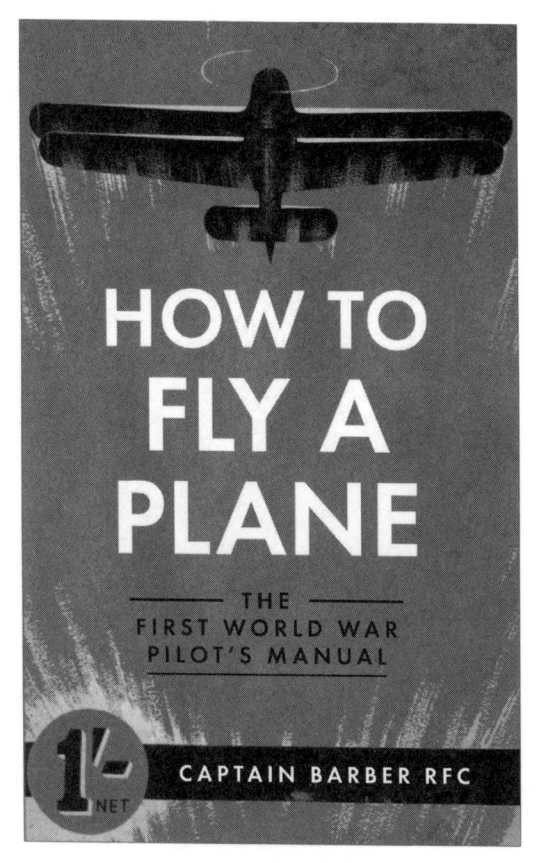

How to Fly a Plane
Captain Barber RFC

The training aid for many a pilot from one of the leading instructors of the First World War.

978 1 4456 3583 5
160 pages

Available from all good bookshops or order direct
from our website www.amberley-books.com

ALSO AVAILABLE FROM AMBERLEY PUBLISHING

How to Pilot a Submarine
US Navy

The training aid for the US submarine fleet at the end of the Second World War.

978 1 4456 3585 9
224 pages

Available from all good bookshops or order direct from our website www.amberley-books.com

ALSO AVAILABLE FROM AMBERLEY PUBLISHING

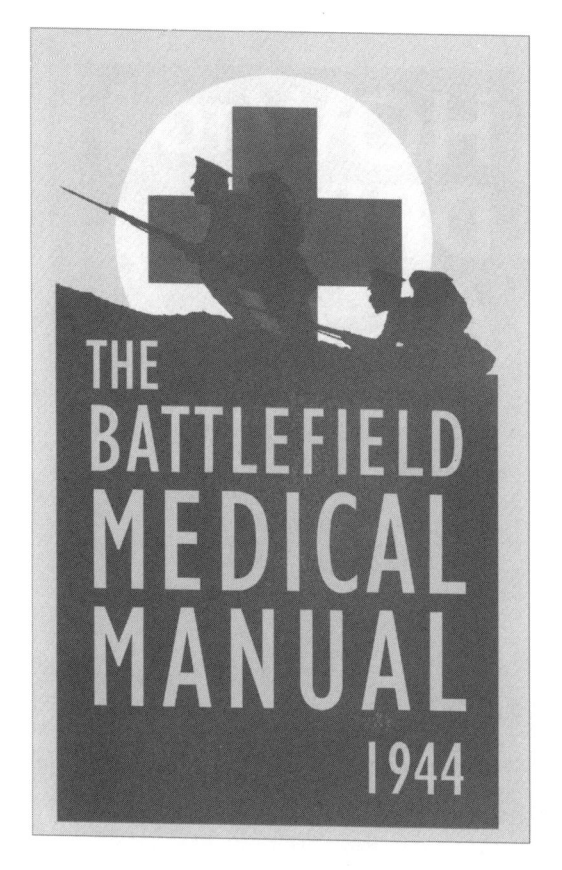

The Battlefield Medical Manual 1944
US Medical Department

In no place is medicine more valued that the battlefield. Giving an intriguing insight into not only field medicine but also living conditions and army training at the time, *The Battlefield Medical Manual* is a book that has, no doubt, saved thousands of lives.

978 1 4456 4312 0

288 pages

Available from all good bookshops or order direct from our website www.amberley-books.com

ALSO AVAILABLE FROM AMBERLEY PUBLISHING

How to Drive a Car
Motor Magazine

A fascinating insight into driving in the 1920s and 1930s.

978 1 4456 3579 8
160 pages

Available from all good bookshops or order direct
from our website www.amberley-books.com

ALSO AVAILABLE FROM AMBERLEY PUBLISHING

Hurricane Manual 1940
Dilip Sarker

The Hawker Hurricane was a vital stalwart of the British war effort.
Here, for the first time, Dilip Sarker collates all the training anuals and
notes on how to fly a Hurricane.

978 1 4456 2120 3
256 pages

Available from all good bookshops or order direct
from our website www.amberley-books.com

ALSO AVAILABLE FROM AMBERLEY PUBLISHING

Spitfire Manual 1940
Dilip Sarker

How to fly the legendary fighter plane in combat using the manuals and instructions supplied by the RAF during the Second World War.

978 1 84868 436 2
288 pages

Available from all good bookshops or order direct from our website www.amberley-books.com

ALSO AVAILABLE FROM AMBERLEY PUBLISHING

Land Girl Manual 1941
W. E. Shewell-Cooper

A fabulous piece of wartime nostalgia, a facsimile edition of the manual used by the Land Girls during the Second World War.

978 1 4456 0279 0
160 pages

Available from all good bookshops or order direct
from our website www.amberleybooks.com

ALSO AVAILABLE FROM AMBERLEY PUBLISHING

Train Driver's Manual
Colin Maggs

Colin Maggs has assembled a fascinating collection of illustrated railwayman's handbooks. You want to be a train driver? This book answers all your questions.

978 1 4456 1680 3
304 pages

Available from all good bookshops or order direct from our website www.amberley-books.com

ALSO AVAILABLE FROM AMBERLEY PUBLISHING

The Home Guard Manual 1941
Edited by Campbell McCutcheon

Over a period of a few months, the rag-tag group known as the Home Guard was armed, uniformed and trained, using the *Home Guard Manual*.

978 1 4456 0047 5
234 pages

Available from all good bookshops or order direct
from our website www.amberley-books.com